Hydro-Environmental Asse
Small Water Bodies:
From Local to Global Scales

小型水体的水环境评估：
从局地到全球尺度

陈文君 贺 斌 著

重庆大学出版社

图书在版编目（CIP）数据

小型水体的水环境评估: 从局地到全球尺度: 英文/
陈文君、贺斌著. —— 重庆: 重庆大学出版社, 2023.8

ISBN 978-7-5689-3840-2

I. ①小... II.①陈... ②贺... III.①区域水环境—环境
质量评价—英文 IV.①X143

中国国家版本馆CIP数据核字（2023）第057585号

小型水体的水环境评估：从局地到全球尺度

陈文君 贺 斌 著

责任编辑：牟 妮 版式设计：牟 妮

责任校对：邹 忌 责任印制：赵 晟

*

重庆大学出版社出版发行

出版人：陈晓阳

社址:重庆市沙坪坝区大学城西路21号

邮编:401331

电话：（023）88617190 88617185（中小学）

传真：（023）88617186 88617166

网址：http://www.cqup.com.cn

邮箱：fxk@cqup.com.cn（营销中心）

全国新华书店经销

重庆新生代彩印技术有限公司

*

开本：720mm×1020mm 1/16 印张：20.25 字数：484千

2023年8月第1版 2023年8月第1次印刷

ISBN 978-7-5689-3840-2 定价：80.00元

Foreword

With the development of our awareness on sustainable development in resources and the environment, many new scientific and practical topics are emerging, such as best management practices, low impact development, and nature-based solutions. Relative to large and well-regarded water bodies, such as lakes and rivers, small water bodies, including both natural and man-made ponds, pools, streams, etc., are sometimes numerous, widely distributed, and tightly linked to socio-economic development, with multiple ecosystem services to the well-being of human society. Their small size and shallow characteristics, however, leave them traditionally unmonitored, unmapped, and poorly protected and less studied in existing environmental geosciences and relative sustainable management practices. Over the last decade, growing attention is paid to small water bodies from the scientific community and management sectors such as environmental protection, sustainable agriculture, urban planning, etc. This new trend brings a set of recent research and practical initiatives that high light the characteristics and importance of small water bodies across different geographical and climatic regions.

This monograph *Hydro-environmental Assessment of Small Water Bodies: From Local to Global Scales* timely focuses on small water bodies, in which the authors put great efforts and invested long time on the hydro-environmental assessment of small water bodies at local scales in Jiangsu Province, southeastern China, and perspectives at regional scale of entire southern China and global scales. Based on multidisciplinary research methods and various datasets, and for farm ponds and small reservoirs of southeastern China and the world's non-floodplain wetlands, the authors have successfully verified the hydro-environmental effects of small water bodies in these regions, and proposed new findings for management and environmental sustainability.

In a nutshell, the results obtained in this monograph substantially enhance the knowledge of hydro-environmental effects of small water bodies, especially farm ponds and small reservoirs, at local and regional scales. This knowledge provides useful means to the sustainable use and protection of small water bodies, including the enhancement of their ecosystem services (particularly the ecosystem services of nutrient retention and flood reduction). Furthermore, the refined models and analytic methods of targeted hydro-environmental assessment are potentially transferable to other regions with abundant small water bodies and applicable in future research. I believe that this monograph will provide a useful knowledge base and tool for researchers and resource managers to improve the sustainable use and protection of small water bodies.

Wenjun Chen and Bin He
May 2023

Acknowledgements

We would like to express our sincere gratitude and appreciation to collaborators from Jinling Institute of Technology, Chinese Academy of Sciences, University of California Merced, and Stockholm University. Also, this book is jointly supported by the National Natural Science Foundation of China [Grant No. 42101476 and 42177065], "Qing Lan" Project and Natural Science Research Project of Higher Education Institutions of Jiangsu Province [Grant No. 21KJB170025], Science and Technology Research Program in Key Areas of Guangdong Province [Grant No. 2020B1111530001 and 2019QN01L682], Natural Science Foundation of Jiangsu Province [Grant No. BK20180115], Initial Startup Funding for the High-level Talents [Grant No. JIT-B-201804] and Research Incubation Project of Jinling Institute of Technology [Grant No. JIT-FHXM-201804], China Postdoctoral Science Foundation [Grant No. 2017M611938], and Postdoctoral Research Funding Programs of Jiangsu Province [Grant No. 1601038B].

Parts of this monograph have been published in the following journal articles:

Wenjun Chen, Josefin Thorslund, Daniel Nover, Mark Rains, Xin Li, Bei Xu, Bin He, Hui Su, Haw Yen, Lei Liu, Huili Yuan1, Jerker Jarsjö, Joshua Viers, 2022. A typological framework of non-floodplain wetlands for global collaborative research and sustainable use.

Wenjun Chen, Daniel Nover, Yongqiu Xia, Guangxin Zhang, Haw Yen, Bin He, 2021. "Assessment of extrinsic and intrinsic influences on water quality variation in subtropical agricultural multipond systems." *Environmental Pollution* 276(19):116689.1-116689.13.

Wenjun Chen, Daniel Nover, Haw Yen, Yongqiu Xia, Bin He, Wei Sun, Joshua Viers, 2020. "Exploring the multiscale hydrologic regulation of multipond systems in a humid agricultural catchment." *Water Research* 184(3):115987.1-115987.18.

Wenjun Chen, Bin He, Daniel Nover, Haiming Lu, Jian Liu, Wei Sun, Wen Chen, 2019. "Farm ponds in southern China: Challenges and solutions for conserving a neglected wetland ecosystem." *Science of the Total Environment* 659: 1322-1334.

Wenjun Chen, Bin He, Daniel Nover, Weili Duan, Kaiyan Zhao, Wen Chen, 2018. "Spatiotemporal patterns and source attribution of nitrogen pollution in a typical headwater agricultural watershed in Southeastern China." *Environmental Science and Pollution Research* 25(3): 2756-2773.

Wenjun Chen, Daniel Nover, Bin He, Huili Yuan, Kaimeng Ding, Jun Yang, Suozhong Chen, 2018. "Analyzing inundation extent in small reservoirs: A combined use of topography, bathymetry and a 3D dam model." *Measurement* 118(9): 202-213.

Wenjun Chen, Bin He, Junting Ma, Chuanhai Wang, 2017. "A WebGIS-based flood control management system for small reservoirs: A case study in the lower reaches of the Yangtze River." *Journal of Hydroinformatics* 19(2): 299-314.

Wenjun Chen, Bin He, Lei Zhang, Daniel Nover, 2016. "Developing an integrated 2D and 3D WebGIS-based platform for effective landslide hazard management." *International Journal of Disaster Risk Reduction* 20: 26-38.

Table of Contents

Chapter 1 Introduction

1.1 Background 1

1.2 Organization of this book 9

References...14

Chapter 2 Challenges and Solutions for Conserving Farm Ponds in Southern China

2.1 Introduction 21

2.2 Farm ponds in southern China 24

2.3 Why are these small and scattered waters important? 27

    2.3.1 Hydrological regulation and ecological cleanup 28

    2.3.2 Biodiversity conservation 29

    2.3.3 Socioeconomic benefits 30

2.4 What are the threats and management challenges? 31

    2.4.1 Inadequate planning and regulations 32

    2.4.2 Rural nonpoint and mini-point source pollution 33

    2.4.3 Climate change 34

    2.4.4 Invasive species 35

    2.4.5 Management challenges 36

2.5 Policies and approaches for farm pond conservation...................38

    2.5.1 Public awareness building 39

    2.5.2 Top-down regulations and bottom-up engagement 40

    2.5.3 Sustainable management and utilization................................42

2.5.4 Inventory mapping 43

2.5.5 IoT-based collaborative monitoring 44

2.5.6 Numerical assessment tools 45

2.6 Conclusions 48

References 49

Chapter 3 Assessment of Nitrogen Pollution in a Pond-rich Agricultural Watershed

3.1 Introduction 57

3.2 Materials and methods 60

3.2.1 Study area and data collection 60

3.2.2 HSPF description and data pre-processing 62

3.2.3 Sensitivity analysis, calibration, and validation 69

3.2.4 Source attribution of N pollution 71

3.3 Results and discussion 72

3.3.1 Calibrated parameters and modeling performance 72

3.3.2 Spatiotemporal analysis and source attributions of N pollution 77

3.3.3 Comparisons with previous studies on N source attribution 80

3.3.4 Implications to headwater agricultural watersheds 84

3.4 Conclusions 85

References 87

Chapter 4 Quantifying the Hydrologic Regulation Effects of Agricultural Multi-pond Systems

4.1 Introduction 94

4.2 Materials and methods 97

4.2.1 Study area and data 97

4.2.2 The SWAT-MPS model 99

4.2.3 Model calibration and validation 104

    4.2.4 Model output and hydrologic analysis 109

4.3 Results 110

    4.3.1 Calibrated parameters and model performance 110

    4.3.2 Pond water balance and MPS identification 113

    4.3.3 MPS' regulation effect on streamflow, baseflow and quickflow 117

    4.3.4 Catchment water balance and extreme hydrographs 120

4.4 Discussion 122

    4.4.1 Multi-scale hydrologic regulation of MPSs 122

    4.4.2 Implications for management 126

    4.4.3 Improved model limitations and recommendations 129

4.5 Conclusions 131

References 132

Appendix 140

Chapter 5 Attributing Influencing Factors of Water Quality Variation in Multi-pond Systems

5.1 Introduction 141

5.2 Materials and methods 145

    5.2.1 Study sites and geospatial data 145

    5.2.2 Sample collection and analysis 146

    5.2.3 Statistical analysis 147

5.3 Results 150

    5.3.1 Spatial and seasonal patterns of water quality 150

    5.3.2 Characteristics of extrinsic and intrinsic factors 152

    5.3.3 Correlation between water quality and influencing factors 155

5.4 Discussion 159

    5.4.1 Extrinsic environments vs. intrinsic characteristics 159

5.4.2 Spatial and seasonal variations of influencing factors......................162

5.4.3 Implications for rural planning and pond management....................164

5.5 Conclusions 166

References..167

Appendix 175

Chapter 6 Syndicating Multi-source Data for Inundation Extent Analysis of Small Reservoirs

6.1 Introduction 183

6.2 Methodology 187

6.2.1 Study area and materials..187

6.2.2 Integrating bathymetry with surrounding topography....................189

6.2.3 Integrating natural terrain with a 3D dam model............................191

6.2.4 Reservoir inundation extent analysis 194

6.3 Results..197

6.3.1 An integrated reservoir terrain model 197

6.3.2 Reservoir inundation extent estimation 198

6.3.3 A comparison between the stage-storage curves 200

6.4 Discussion 201

6.5 Conclusion...205

References..205

Chapter 7 Developing a WebGIS-based Flood Control Management System for Small Reservoirs

7.1 Introduction 211

7.2 Methodology 214

7.2.1 Study area and materials..214

7.2.2 Adopting agile methods to involve different participants.................215

7.2.3 System architecture design..217

7.2.4 Development of the flood level forecasting model.........................218

7.2.5 Development of reservoir maps..222

7.3 Results and discussion..223

7.3.1 Model parameter acquisition and verification............................223

7.3.2 Application to flood control management...................................226

7.3.3 Discussion..227

7.4 Conclusions...230

References...231

Chapter 8 Developing an Integrated 2D & 3D WebGIS Platform for Scattered Points of Interest

8.1 Introduction...235

8.2 Study area..238

8.3 Methodology...240

8.3.1 Requirement analysis...240

8.3.2 Architecture design..242

8.3.3 Modules design...244

8.3.4 Algorithms for 2D and 3D dynamic display.................................248

8.4 Results and discussion..252

8.4.1 Landslide multi-level management...252

8.4.2 Landslide emergency response...254

8.4.3 Prevention information for public access.................................255

8.4.4 Discussion...256

8.5 Conclusion..259

References...260

Chapter 9 Exploring Global Collaboration Opportunities for Non-floodplain Wetland Conservation

9.1 Introduction 265

9.2 NFWs have been widely studied with strong North America biases.........267

9.3 Emerging research patterns across representative NFW regions 269

9.4 New opportunities for global coordinated management efforts 272

9.5 Connecting the disconnected..277

References...278

Appendix 284

Chapter

1

Introduction

1.1 Background

Small water bodies, which in this monograph refer to ponds, pools, ditches, springs, and small reservoirs largely located in headwater regions, are numerous and widely distributed across the world (Fig. 1.1). Our awareness of the importance of small water bodies has witnessed an increasing growth since the last decade (Biggs et al., 2017; Calhoun et al., 2017; Chen et al., 2019), due to their abundance, importance to freshwater systems and biodiversity, contributions to ecosystem services, and their vulnerability and sensitivity to a variety of human activities (Cohen et al., 2016; Creed et al., 2017; Golden et al., 2019; Sayer and Greaves, 2020; Lane et al., 2022). Meanwhile, our lack of knowledge on the functional services of these small water bodies is widely recognized, which affects practical and effectively legislative and regulative measures to protect and sustainable use of these waters (Biggs et al., 2005; Sullivan et al., 2019; Wade et al., 2022). This circumstance has led to many recent local and regional initiatives that highlight the importance of small water bodies and encourage collaborations between scientists, decision makers, local planners, travelers, and other interested parties who pursue their better management and utilization (Takeuchi et al., 2016; Xu et al., 2019; Wang et al., 2019), especially in the global trend of climate change, pandemic diseases and food insecurity (Duan et al., 2016; Thorslund et al., 2017; Chen et al.,

2022). Focused on these novel important academic and management topics, our monograph aims to provide a series of hydro-environmental assessment of small water bodies from local to global scales. These assessments include nitrogen and non-point source pollution (water quality), hydrologic regulation effects (water resource), and inundation analysis and flood control management, as well as research perspectives at two scales of southern China and the world. As a pilot study of hydro-environmental assessment of small water bodies, this monograph can provide monitoring strategies and modeling tools for regions with abundant small water bodies and similar concerns on these waters. It can also provide reference to water resource management, water environment protection, regional planning and development, sustainable agriculture, and other areas related with small water bodies.

Fig. 1.1 Examples of small water bodies (Source: Tiner, 2003). From top to bottom and left to right, these examples are playas, ponds, wetlands, small lakes, pools, and seepage water bodies.

Deterioration of water quality, especially nitrogen and phosphorus, has attracted growing attention across the globe. The excessive normal nutrient and other dissolved substances including persistent organic pollutants and antibiotics have along accumulated in various landforms and flowed into the water environment along with rainfall runoff processes (Lane et al., 2018; Golden et al., 2019). Previous studies largely focused on large rivers, lakes and coasts, with fewer attention given to small water bodies in headwater regions (Biggs et al., 2017; Chen et al., 2018; Sayer et al., 2020). The neglect of these water bodies, which often serve as nutrient storage and retention in the watersheds, hinders our deeper understanding of nutrient accumulation processes and associated pollution control projects. This is because that small water bodies are usually interacted with surface, shallow subsurface and groundwater flows in the aquatic systems (Lane et al., 2018; Chen et al., 2020), and they are a mediator between upstream landforms and downstream waters. Early in the 1990s, scientists suggested that small water bodies had potential impact on watershed nutrient conditions using landscape-scale statistical methods (Whigham et al., 1988; Johnston et al., 1990). In the early 2000s, hydrological processes of small water bodies embedded in watershed were explored, providing glimpses of these waters in the timing and frequency of their water transport to downgradient rivers and lakes (Leibowitzet al., 2003; Richardson, 2003; Rains et al., 2006). Later in the 2010s, cumulative hydrological effects were further demonstrated in watershed-scales, but few studies directly targeted at small water bodies and their individual or grouped effects on downstream water quality (Lane et al., 2018; Golden et al., 2019; Chen et al., 2021). Although intrinsic characteristics of small water bodies' (e.g., water storage, connectivity, and bidirectional exchange with shallow groundwater; Xia et al., 2016; Chen et al., 2018) and extrinsic factors of subcatchment (e.g., slope, land use types, and soil attributes; Li et al., 2019; Wang et al., 2018) and precipitation (e.g., intensity, duration, and concentration; Zhang et al., 2020) were considered as water quality influences (Fig. 1.2), fragmentary knowledge on small water bodies and their cumulative watershed-scale effects remains a challenge for both researchers and managers.

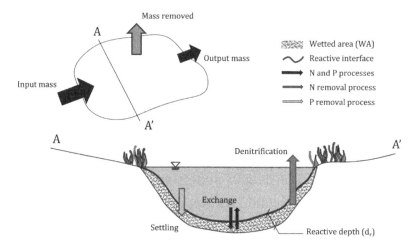

Fig. 1.2 Conceptual model and associated biogeochemical processes of nutrient retention and removal by small water bodies (Source: Cheng and Basu, 2017)

More essentially, small water bodies are landscape functional elements performing physical and hydrological effects on downstream systems (Marton et al., 2015; Rains et al., 2015; Schofield et al., 2018). Source, sink and refuge are the three main functions first identified for these waters (Leibowitz et al., 2008). Later in 2015, USEPA added another two functions, lag and transformation, into this conceptual framework. Such framework helps to study and understand the physical and hydrological connections between small water bodies and their downstream and adjacent water systems (Leibowitz et al., 2018; Schofield et al., 2018). Based on these insights, the interdisciplinary hydrological science community has started looking into several research subjects of small water bodies in various landscapes, such as spatial heterogeneity, physical interactions, and connectivity (Fig. 1.3). The area scale of studied watersheds ranges from a few hectares to several thousand square kilometers (Evenson et al., 2016; Evenson et al., 2018; Yeo et al., 2019; Chen et al., 2020; Zeng and Chu, 2021). However, due to the different intrinsic characteristics and extrinsic factors of small water bodies across the world, physical processes and related driving mechanisms are still not clear, especially involving various management and conservation measures (Calhoun et al., 2017; Lahoz-Monfort et al., 2019). On the other hand, more than 847,000 reservoirs have been constructed globally in the last 100 years, and approximately 95% of them are small reservoirs (Song et al. 2015), not mentioning those unmanaged small reservoirs. They provide water, irrigation, and sometimes hydroelectric energy, and more importantly, stabilize extreme inflows to mitigate floods or droughts (Chang, 2006; Rodrigues et al., 2012). Comparing with natural small water bodies, more attention is paid to water resource and flood control management for small reservoirs, including simultaneous considerations on the hydrological, geotechnical, and environmental aspects (Chen et al., 2017). Although a number of countries have employed flood forecasting systems, considering volume, peaks and duration of rainfall events, since the 1990s, their application and usage require a deeper understanding of flood processes and a more effective operational solution, especially for small reservoirs with a fast process of converging and rising of water levels (Werner et al., 2013; Cools et al., 2016).

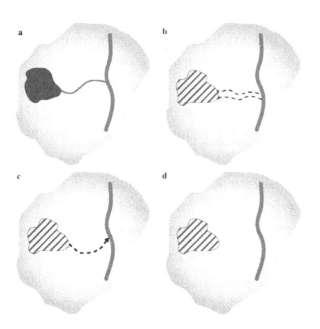

Fig. 1.3
Four types of hydrological connections between small water bodies and streams or rivers (Source: Lane et al., 2018). A water body connected to a river by surface flow via (a) a headwater stream and (b) noncanalized marshes, and (c) connected to a river by groundwater flow, as well as (d) a water body that is hydrologically isolated from a river.

Our mapping, measuring and modeling techniques for small water bodies have gradually developed since the last decade, but are still limited compared with those for rivers, coasts, and lakes (Golden et al., 2019; Swartz and Miller, 2021). Scientists and decision-makers need to acquire where small water bodies locate before their hydro-environmental assessment. In addition to traditional satellite observation and field visit, including setting up ground instruments like cameras, water level meters, etc., Light Detection and Ranging (LiDAR) and novel radar-based remote sensing methods are started in detecting small water bodies, especially in forests and canopy environments (Lang and McCarty, 2009; Tiner et al., 2015; Wu et al., 2019). Many wetland inventories and lake database are available online, such as HydroLAKES and MERIT Hydro. These datasets, however, mostly don't have enough spatial resolution for capturing and assessing small water bodies, especially within changing landscapes (Tiner et al., 2015; DiBello et al., 2016). Accordingly, our measurements with high temporal resolution for the nutrient levels and hydrological regimes of small water bodies are lacking. This has been called on for more than 15 years, although it is resource intensive and challenge to measure those variables in watershed scales (Whigham and Jordan, 2003; Biggs et al., 2005; Chen et al., 2019). On the other hand, models, especially process-based models, are useful tools to assess how small water bodies affect hydrological processes and nutrient loads in watersheds. Recent model advances have enabled Soil and Water Assessment Tool (SWAT), Hydrological Simulation Program-Fortran (HSPF), and Annualized Agricultural Non-Point Source Pollution Model (AnnAGNPS), to incorporate small, scattered wetlands (Golden et al., 2017; Chen et al., 2018; Yasarer et al., 2018). The spatially-explicit representation with physical and functional attributes provides feasible means for detailed assessments on these waters (Evenson et al., 2015; Yasarer et al., 2018; Evenson et al., 2018; Yeo et al., 2019). However, model limitations still exist for consideration of MPSs, including bidirectional exchanges between ponded water and shallow groundwater (Rains et al., 2015), and cascading fill-spill relationships between intermittently connected ponds (Yu et al., 2015). These model deficiencies and lack in mapping and measuring together hinder the detailed assessment of small water bodies.

From a typological perspective, small water bodies can be ponds, pools, bogs, fens, marshes, playas, and others across different parts of the world (Mushet et al., 2015; Creed et al., 2017; Chen et al., 2022). Scientists and decision-makers hold various research interests among these small water body types, as they are closely associated with our life and daily living environment in different ways (Biggs et al., 2017; Hunter et al., 2017; Calhoun et al., 2017; Thorslund et al., 2017; Ghajarnia et al., 2020). In southern China, the farm pond, whose history can be dated back to 700s BCE, is a widely distributed type of small water body. Farm ponds were first constructed in Anhui Province, and were vigorously developed and flourished as a unique natural-economic-social complex across southern China since 1949 (Verhoeven et al., 2006; Liu et al., 2009; Yu et al., 2015). Some larger ponds were later enlarged as small reservoirs, due to appropriate topographic features of the surrounding landscape and hydrologic and vegetation conditions, so that they can hold and manage greater water resources (Yu, 2015). In conjunction with the surrounding woods, farmlands, cottages, and water courses, farm ponds and small reservoirs constitute a unique pastoral landscape in southern China (Fig. 1.4). These small water bodies with Chinese characteristics can alter the hydrological and nutrient cycling in the watershed, and lead to idiosyncratic riparian conditions to host flora and fauna. They also offer considerable social and cultural benefits, including improving physical and mental well-being and increasing awareness of environmental conservation at larger scales (Yu et al., 2015; Creed et al., 2017; Chen et al., 2022). For example, the 193 farm ponds can reduce annual irrigation water shortages from 306±26 to 89±48 mm in the Liuchahe catchment, Anhui Province (Yin et al., 2006), while reducing the flood peak from 2.5 to 0.3 m^3 s^{-1} during a heavy rainfall event (141 mm d^{-1}; Liu et al., 2009). Although hydrologic regulation of these small water bodies has been gradually recognized in catchments (Wang et al., 2008; Longbucco, 2010; Feng et al., 2013), the comprehensive hydro-environmental assessment relating with regional planning, environmental protection, sustainable agriculture, etc., is still a multidisciplinary challenging issue, and can serve as a testbed for other types of small water bodies across the globe.

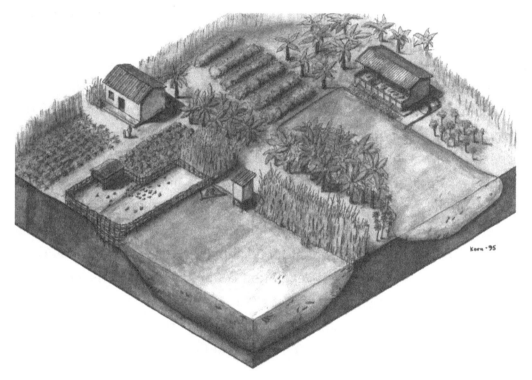

Fig. 1.4 Artist view of the complete integrated pond farming concept in southern China (Source: Korn, 1996)

To sum up, global changes including all the future uncertainties due to human activities have significantly affected hydrological cycles and associated water resources, nutrient behaviors, and biological responses, especially during extreme rainfall and drought events. The widely distributed small water bodies, including various types around the world, play non-negligible roles in these watershed processes, but our understandings are lacking due to limited map resolution, coverage of measurement, and targeted modeling tools. Farm ponds and small reservoirs, as a a unique and pastoral landscape in southern China, can serve as a testbed for hydro-environmental assessment of other types of small water bodies. Hence, the objective of this book is to assess the hydro-environmental processes of farm ponds and small reservoirs in local regions of southern China, and propose research perspectives that inform both farm ponds and other small water body types around the world.

1.2 Organization of this book

Our hydro-environmental assessment of small water bodies covers local to global scales. The research topic started from a typical agricultural watershed (the Chenzhuang Watershed) and a small reservoir (the Hengshan Reservoir) in Jiangsu Province, southeastern China. It then expanded to all farm ponds in southern China, and non-floodplain wetlands (i.e., a common type of wetlands, surrounded by uplands outside of floodplains and riparian areas; Mushet et al., 2015) across the globe. To make the story line more understandable, Chapter 2 first introduces farm ponds at the scale of southern China. Chapter 3, 4 and 5 then study hydrological processes and nitrogen pollution in the agricultural watershed with multi-pond systems. Chapter 6 and 7 proposes computational geometry and WebGIS-based technologies to depict inundation in small reservoirs and associated flood control management. At last, Chapter 8 expands these studies to a global scale and proposes new insights into global collaboration opportunities for non-floodplain wetland conservation.

Chapter 1 depicts the background and concludes the research progress of hydro-environmental assessment of small water bodies, suggesting farm ponds is one of the typical small water bodies among global-extensively studied small vulnerable waters.

Chapter 2 introduces the hydrologic, biogeochemical, and socioeconomic benefits provided by farm ponds in southern China for thousands of years, and their contemporary threats and management challenges, including (1) inadequate planning in terms of construction and conservation regulations; (2) rural nonpoint source and mini-point source pollution; (3) climate change induced abnormalities in the hydroperiod and disturbance to wildlife; (4) invasive species; and (5) inadequate social and political capacity to consider ecological conservation. To conserve such neglected wetland ecosystem, this chapter highlights approaches that build public awareness and involve inventory maps as a basis. Policies that integrate top-down regulation and bottom-up engagement and emphasize sustainable management and utilization are recommended to ensure the effectiveness and continuous improvement of conservation programs. Techniques that involve interconnected smart sensors, volunteering and citizen science, and integrated process-based modeling are preferred when conducting comprehensive descriptions of the pond landscape, numerical assessments on their ecosystem services, and associated conservation cost analyses. The analytical framework and conservation suggestions are referential to sustainable rural development and the management of other small, scattered wetlands.

Chapter 3 studies nitrogen pollution and source attribution for a multi-pond system around a village using the HSPF model, taking a typical small watershed in the low mountains of southeastern China as an example. This chapter exhibits distinctive spatial seasonal variations with an overall seriousness rank for the three indicators: total nitrogen (TN) > nitrate/nitrite nitrogen (NO_x^--N) > ammonia nitrogen (NH_3-N), according to the Chinese Surface Water Quality Standard. TN pollution was severe for the entire watershed, while NO_x^--N pollution was significant for ponds and ditches far from the village, and the NH_3-N concentrations were acceptable except for the ponds near the village in summer. Although food and cash crop production accounted for the largest source of N loads, mini-point pollution sources, including animal feeding operations, rural residential sewage and waste, together contributed as high as 47% of the TN and NH_3-N loads in ponds and ditches. So, apart from eco-fertilizer programs and concentrated animal feeding operations, the importance of environmental awareness building for resource management is highlighted for small farmers in headwater agricultural watersheds.

Chapter 4 presents a novel approach to quantify the multi-scale hydrologic regulation of multi-pond systems (MPSs), a common type of small, scattered wetland in humid agricultural regions. A new version of Soil and Water Assessment Tool (SWAT) was developed to incorporate improved representation of: (1) perennial or intermittent spillage connections of pond-to-pond and pond-to-stream, and (2) bidirectional exchange between pond surface water and shallow groundwater. SWAT-MPS, which adopts rule-based artificial intelligence to model the possibilities of different spillage directions and GA-based parameter optimization over the two simulation years (June 2017 to May 2019), successfully replicated streamflow and pond water-level variations in a 4.8 km^2 test catchment, southern China. Water balance analysis and scenario simulations were then executed to assess the hydrologic regulation at single pond, single MPS, and entire catchment scales. Results revealed (1) the presence of 9 series- or series-parallel connected MPSs, in which pond overflow accounted for as much as 59% of the catchment water yield; (2) seasonally- and MPS-independent baseflow support and quick flow attenuation, with ranked level of pond water storage for baseflow support across different land-use types: forest > farm > village, and inversed correlation of pond spillage to baseflow and quick flow variations in the farmland; and (3) MPS-aggregated catchment flood peak reduction (> 20%) and baseflow increment (26%) in the following dry days.

Chapter 5 employs a combined self-organizing map (SOM) and a linear mixed-effects model (LMEM) to relate water quality variation of multi-pond systems (MPSs) to their extrinsic and intrinsic influences. Across the 6 test MPSs with environmental gradients, ammonium nitrogen (NH$_4^+$-N), total nitrogen (TN), and total phosphate (TP) almost always exceeded the surface water quality standard (2.0, 2.0, and 0.4 mg/L, respectively) in the up- and mid-stream ponds, while chlorophyll-a (Chl-a) exhibited a hypertrophic state ($\geqslant$ 28 μg/L) in the midstream ponds during the wet season. Synergistic influences explained 69±12% and 73±10% of the water quality variations in the wet and dry season, respectively. The adverse, extrinsic influences were generally 1.4, 6.9, 3.2, and 4.3 times of the beneficial, intrinsic influences for NH_4^+-N, nitrate nitrogen (NO_3^--N), TP, and potassium permanganate index (COD_{Mn}), respectively, although the influencing direction and degree of forest and water area proportion were spatiotemporally unstable. While COD_{Mn} was primarily linked with rural residential areas in the midstream, higher TN and TP concentrations in the up- and mid-stream were associated with agricultural land, and NH_4^+-N reflected a small but non-negligible source of free-range poultry feeding. Pond surface sediments exhibited consistent, adverse effects with amplifications during rainfall, while macrophyte biomass can reflect the biological uptake of COD_{Mn} and Chl-a, especially in the mid- and down-stream during the wet season.

Chapter 6 focuses on the combined use of various common tools and GIS techniques to achieve rapid and effective estimation of inundation extent for both small reservoirs and associated hydraulic structures. It revealed that the reservoir terrain surface can be more realistic and coherent, if an airborne LiDAR-derived DEM, bathymetric measurements, and a 3D dam model are integrated. The source flooding algorithm based on a horizontal water surface can be enhanced by introducing the intersection detection from computational geometry, so that submerged areas could spread from terrain mesh cells to the upstream face of embankment dams, although the required neighbouring relationship is absent in the 3D geometric model. Taking the Hengshan Reservoir in southeastern China as the study area, the proposed methods are tested by a historical rainfall event during the typhoon season. They are also proved to support extreme water stages, and better-founded to determine the reservoir stage-storage relationship, which is mostly based on bathymetric measurements and a polynomial extrapolation.

Chapter 7 proposes a WebGIS-based flood control management system to support the flood discharge of small reservoirs during intensive rainfall in the flood season. The agile software development method and a loosely coupled structure are used to combine multidisciplinary knowledge from different experts. A flood level forecasting model for reservoirs in humid regions is established based on rainfall and water level measurements. This chapter aims to provide concise information for reservoir managers to choose an appropriate discharging scheme, so that the capacity is maintained in a safe range on the next day. Using the Hengshan Reservoir in the lower reaches of the Yangtze River as an example, the model verification reveals that it is acceptable for rainfall events whose daily amount is near or above 100 mm (the heavy rainstorm level in China), and the system is verified by a trial operation during the typhoon season. While most existing flood control systems focus on river basins and large reservoirs, this chapter considers the data availability and practical flood discharging scenario of small reservoirs, and provides a useful tool for the flood control management.

Chapter 8 presents a novel 2D and 3D WebGIS-based platform for the management of small, scattered points of interest. These points could be small water bodies and other natural resources, and here, they are landslide hazards, while the management aims at landslide multi-level management and emergency response. The scalable network architecture and three-tier software architecture are designed to support survey data improvement performed by geological surveys in different administrative levels, fast spatial decision support for rescue and evacuation after sudden hazard incidents, as well as prevention information for public access. The web service integration is widely applied in the platform, and proved useful for keeping landslide-related information consistent and up-to-date, since it relies on joint efforts from different government departments with expertise, rather than the local and independent storage pattern. The combined utilization of 2D and 3D WebGIS takes advantage of their respective superiorities, and generates a superior display and analytic web environment for local decision-makers. After elaborating the main modules and key algorithms, this user-friendly platform has been experimented and accepted by three levels of geological surveys in Zhejiang Province, China, and presented as an integrated WebGIS environment for effective management of small, scattered points of interest, such as landslide hazards, in large areas.

Chapter 9 synthesizes recent work to explore how extensively non-flood wetlands (NFWs) have been studied and possible patterns of their research focuses around the world, and discuss new opportunities for sustainable NFW use for the global community. Results indicate that NFWs have been widely studied across all continents, but have strong North America biases in the literature. We hypothesize such biases stem from terminology rather than real geographical bias around existence and functionality. To confirm this, this chapter then explores a set of geographically representative NFW regions around the world, the characteristics of research focuses for these regions, and the monetized ecosystem services of NFWs in the world's major economies. The identified commonalities and patterns suggest that there is more that unites disparate wetland research and management efforts than we might otherwise appreciate. They also support opportunities for learning and collaboration, if we can move beyond terminology and focus on wetland functionality. Based on these findings, this chapter outlines four pathways that will aid in reaching better collaborative actions across scales, including classification and simulation in theoretical and technical aspects, and improved legislative support and educational measures.

References

Biggs, J., Von, F. S., Kelly, Q. M., et al., 2017. "The importance of small waterbodies for biodiversity and ecosystem services: Implications for policy makers". *Hydrobiologia* 793(1):3-39.

Biggs, J., Von, F. S., Kelly, Q. M., et al., 2005. "15 years of pond assessment in Britain: Results and lessons learned from the work of pond conservation." *Aquatic Conservation: Marine and Freshwater Ecosystems* 15(6):693-714.

Calhoun, A. J., Mushet, D. M., Bell, K. P., et al., 2017. "Temporary wetlands: Challenges and solutions to conserving a 'disappearing' ecosystem." *Biological Conservation* 211(2):3-11.

Chang, F., and Chang, Y., 2006. "Adaptive neuro-fuzzy inference system for prediction of water level in reservoir." *Advances in Water Resources* 29 (1):1-10.

Cheng, F.Y., and Basu, N.B., 2017. "Biogeochemical hotspots: Role of small water bodies in landscape nutrient processing." *Water Resources Research* 53(6):5038-5056.

Chen, W., He, B., Nover, D., et al., 2019. "Farm ponds in southern China: Challenges and solutions for conserving a neglected wetland ecosystem." *The Science of the Total Environment* 659:1322-1334.

Chen, W., He, B., Nover, D., et al., 2018. "Spatiotemporal patterns and source attribution of nitrogen pollution in a typical headwater agricultural watershed in Southeastern China." *Environmental Science & Pollution Research* 25(3):2756-2773.

Chen, W., Nover, D., Yen, H., et al., 2020. "Exploring the multi-scale hydrologic regulation of multi-pond systems in a humid agricultural catchment." *Water Research* 184(3):115987.

Chen, W. J., Daniel, N., Haw, Y., et al., 2020. "Exploring the multi-scale hydrologic regulation of multi-pond systems in a humid agricultural catchment." *Water Research* 184(3): 115987.

Chen, W. J., Daniel, N., Xia, Y. Q., et al., 2021. "Assessment of extrinsic and intrinsic influences on water quality variation in subtropical agricultural multi-pond systems." *Environmental Pollution* 276(19): 116689.

Cohen, M. J., Creed, I. F., Alexander, L., et al., 2016. "Do geographically isolated wetlands influence landscape functions?" *Proceedings of the National Academy of Sciences of the United States of America* 113:1978-1986.

Cools, J., Innocenti, D., O'Brien, S., 2016. "Lessons from flood early warning systems." *Environmental Science & Policy* 58:117-122.

Creed, I. F., Lane, C. R., Serran, J. N., et al., 2017. "Enhancing protection for vulnerable waters." *Nature Geoscience* 10(11): 809-815.

DiBello, F.J., Calhoun, A.J., Morgan, D.E., et al., 2016. "Efficiency and detection accuracy using print and digital stereo aerial photography for remotely mapping vernal pools in New England landscapes." *Wetlands.* 36(3): 505-514.

Duan, W., He, B., Nover, D., et al., 2016. "Floods and associated socioeconomic damages in China over the last century." *Natural Hazards* 82(1): 401-413.

Evenson, G. R., Golden, H. E., Lane, C. R., et al., 2016. "An improved representation of geographically isolated wetlands in a watershed-scale hydrologic model." *Hydrological Processes* 30(22):4168-4184.

Evenson, G. R., Jones, C. N., McLaughlin, D. L., et al., 2018. "A watershed-scale model for depressional wetland-rich landscapes." *Journal of Hydrology* 1:100002.

Feng, X. Q., Zhang, G. X., Jun, X. Y., 2013. "Simulation of hydrological processes in the Zhalong wetland within a river basin, Northeast China." *Hydrology and Earth System Sciences* 17(7):2797-2807.

Ghajarnia, N., Destouni, G., Thorslund, J., et al., 2020. "Data for wetlandscapes and their changes around the world Earth." *Earth System Science Data* 12(2):1083-1100.

Golden, H. E., Rajib, A., Lane, C. R., et al., 2019. "Non-floodplain wetlands affect watershed nutrient dynamics: A critical review" *Environmental Science and Technology* 53(13):7203-7214.

Hunter, J., Acuña, V., Bauer, D. M., 2017. "Conserving small natural features with large ecological roles: A synthetic overview." *Biological Conservation* 211(2):88-95.

Johnston, C. A., Detenbeck, N. E., Niemi, G. J., 1990. "The cumulative effect of wetlands on stream water quality and quantity. A landscape approach." *Biogeochemistry* 10 (2):105-141.

Korn, M., 1996. "The dike-pond concept: Sustainable agriculture and nutrient recycling in China." *Ambio* 25(1): 6-13.

Lahoz-Monfort, J.J., Chadès, I., Davies, A., et al., 2019. "A call for international leadership and coordination to realize the potential of conservation technology." *Bioscience* 69(10): 823-832.

Lane, C. R., Creed, I. F., Golden, H. E., et al., 2022. "Vulnerable waters are essential to watershed resilience." *Ecosystems* 1-28.

Lane, C. R., Leibowitz, S. G., Autrey, B. C., et al., 2018. "Hydrological, physical, and chemical functions and connectivity of non-floodplain wetlands to downstream waters: A review." *Journal of the American Water Resources Association* 54(2):346-371.

Lang, M. W., and McCarty, G. W., 2009. "LIDAR intensity for improved detection of inundation below the forest canopy." *Wetlands.* 29 (4):1166–1178.

Leibowitz, S. G., and Nadeau, T. L., 2003. "Isolated wetlands: State-of-the-science and future directions." *Wetlands.* 23(3):663-684.

Leibowitz, S. G., Wigington, P. J., Rains, M. C., et al., 2008. "Non-Navigable Streams and Adjacent Wetlands: Addressing Science Needs Following the Supreme Court's Rapanos Decision." *Frontiers in Ecology and the Environment* 6(7):366-373.

Leibowitz, S. G., Wigington, Jr., Schofield, K., et al., 2018. "Connectivity of Streams and Wetlands to Downstream Waters: An Integrated Systems Framework." *Journal of the American Water Resources Association* 54 (2): 298-322.

Liu, Y., Fu, Q., Yin, C., 2009. "Phosphorus sorption and sedimentation in a multi-pond system within a headstream agricultural watershed." *Water Quality Research Journal of Canada* 44(3):243-252.

Li, D., Zheng, B., Chu, Z., et al., 2019. "Seasonal variations of performance and operation in field-scale storing multi-pond constructed wetlands for nonpoint source pollution mitigation in a plateau lake basin." *Bioresource Technology.* 280:295-302.

Longbucco, N., 2010. "Impacts of small water bodies on the hydrological response of small agricultural watersheds." Southern Illinois University, Carbondale, Illinois.

Marton, J. M., Creed, I. F., Lewis, D., et al., 2015. "Geographically Isolated

Wetlands are Important Biogeochemical Reactors on the Landscape." *Bioscience* 65(4):408-418.

Mushet. D. M., Calhoun, A. J., Alexander, L. C., et al., 2015. "Geographically isolated wetlands: Rethinking a misnomer." *Wetlands.* 35(3):423-431.

Rains, C. M., Fogg, G. E., Harter, T., et al., 2006. "The role of perched aquifers in hydrological connectivity and biogeochemical processes in vernal pool landscapes, Central Valley." *Hydrological Processes* 20(5):1157-1175.

Rains, M.C., Leibowitz, S. G., Cohen, M. J., et al., 2015. "Geographically Isolated Wetlands Are Part of the Hydrological Landscape." *Hydrological Processes* 30 (1):153-160.

Richardson, C. J., 2003. "Pocosins: Hydrologically isolated or integrated wetlands on the landscape?" *Wetlands.* 23(3):563–576.

Rodrigues, L. N., Sano, E. E., Steenhuis, T. S., et al., 2012. "Estimation of small reservoir storage capacities with remote sensing in the Brazilian Savannah region." *Water Resources Management* 26(4):873-882.

Sayer, C. D., and Greaves, H. M., 2020. "Making an impact on UK farmland pond conservation." *Aquatic Conservation Marine and Freshwater Ecosystems* 30(9):1821-1828.

Schofield, K. A., Alexander, L. C., Ridley, C. E., et al., 2018. "Biota Connect Aquatic Habitats Throughout Freshwater Ecosystem Mosaics." *Journal of the American Water Resources Association* 54(2): 372-399.

Song, W., Jiang, Y., Lei, X., et al., 2015. "Annual runoff and flood regime trend analysis and the relation with reservoirs in the Sanchahe River Basin, China." *Quaternary International* 380(4):197-206.

Sullivan, S., Rains, M. C., Rodewald, A. D., 2019. "Opinion: The proposed change to the definition of 'waters of the United States' flouts sound science." *Proceedings of the National Academy of Sciences* 116(24):11558-11561.

Swartz, T. M., and Miller, J. R., 2021. "The American pond belt: An untold story of conservation challenges and opportunities." *Front. Ecol. Environ.* 19(9):501-509.

Takeuchi, K., Ichikawa, K., Elmqvist, T., 2016. "Satoyama landscape as social-ecological system: Historical changes and future perspective." *Current opinion in environmental sustainability* 19:30-39.

Thorslund, J., Jarsjo, J., Jaramillo, F., et al., 2017. "Wetlands as large-scale nature-based solutions: Status and challenges for research, engineering and management." *Ecological Engineering*108:489-497.

Tiner R. W., 2003. "Geographically isolated wetlands of the United States." *Wetlands.* 23(3):494-516.

Tiner, R.W., Lang, M.W., Klemas, V.V., 2015. *Remote Sensing of Wetlands: Applications and Advances.* Boca Raton: CRC Press.

Verhoeven, J. T., Arheimer, B., Yin, C., et al., 2006. "Regional and global concerns over wetlands and water quality." *Trends in Ecology & Evolution* 21(2):96-103.

Wang, L., Tong, J., Li, Y., 2019. "River chief system (RCS): An experiment on cross-sectoral coordination of watershed governance." *Frontiers of Environmental Science & Engineering* 13(4):1-3.

Wang, X., Yang, W., Melesse, A. M., 2008. "Using hydrologic equivalent wetland concept within SWAT to estimate streamflow in watersheds with numerous wetlands." *Transactions of the American Society of Agricultural and Biological Engineers* 51(1):55-72.

Wang, X., and Zhang, F., 2018. "Multi-scale analysis of the relationship between landscape patterns and a water quality index (WQI) based on a stepwise linear regression (SLR) and geographically weighted regression (GWR) in the Ebinur Lake oasis." *Environmental Science and Pollution Research* 25(7):7033-7048.

Wen, J., Chen, B. H., Junting, M., et al., 2017. "A WebGIS-based flood control management system for small reservoirs: A case study in the lower reaches of the Yangtze River." *Journal of Hydroinformatics* 19(2):299-314.

Werner, M., Schellekens, J., Gijsbers, P., et al., 2013. "The Delft-FEWS flow forecasting system." *Environmental Modelling & Software* 40: 65-77.

Whigham, D. F.; Chitterling, C.; Palmer, B., 1988. "Impacts of freshwater wetlands on water quality: A landscape perspective." *Journal of*

Environmental Management 12 (5):663-671.

Whigham, D. F., and Jordan, T. E., 2003. "Isolated wetlands and water quality." *Wetlands.* 23 (3):541−549.

Wu, Q., Lane, C. R., Li, X., et al., 2019. "Integrating LiDAR data and multi-temporal aerial imagery to map wetland inundation dynamics using Google Earth Engine." *Remote Sensing of Environment* 228:1-13

Xia, Y., Ti, C., She, D., et al., 2016. "Linking river nutrient concentrations to land use and rainfall in a paddy agriculture-urban area gradient watershed in southeast China." *Science of the Total Environment* 566:1094-1105.

Xu, W., Fan, X., Ma, J., et al., 2019. "Hidden loss of wetlands in China." *Current Biology* 29 (18):3065-3071.

Yasarer, L. M., Bingner, R. L., Momm, H. G., 2018. "Characterizing ponds in a watershed simulation and evaluating their influence on streamflow in a Mississippi watershed." *Hydrological Sciences Journal* 63(2):302-311.

Yeo, I. Y., Lee, S., Lang, M. W., et al., 2019. "Mapping landscape-level hydrological connectivity of headwater wetlands to downstream waters: A catchment modeling approach-Part 2." *Science of the Total Environment* 653:1557-1570

Yin, C., Shan, B., Mao, Z., 2006. "Sustainable water management by using wetlands in catchments with intensive land use." In: *Wetlands and natural resource management, Ecological Studies*, vol 190. Berlin: Springer.

Yu, K.J., Jiang, Q.Z., Wang, Z.F., et al., 2015. "The Research progress and prospect of Beitang landscape." *Areal Research and Development* 34(3):130-136.

Zeng, L., and Chu, X., 2021. "Integrating depression storages and their spatial distribution in watershed-scale hydrologic modeling." *Advances in Water Resources* 151(8):103911.

Zhang, W., Li, H., Pueppke, S. G., et al., 2020. "Nutrient loss is sensitive to land cover changes and slope gradients of agricultural hillsides: Evidence from four contrasting pond systems in a hilly catchment." *Agricultural Water Management* 237(12):106165.

Chapter
2

Challenges and Solutions for Conserving Farm Ponds in Southern China

2.1 Introduction

Farm ponds are small (1 m^2 to approximately 0.05 km^2), shallow (< 8 m), natural or man-made surface depressions that permanently or temporarily hold water. They are frequently seen in agricultural regions in many countries (Yin and Shan, 2001; Céréghino et al., 2014; Grobicki et al., 2016). Owing to the low construction cost, high irrigation and drainage efficiency, and value for multiple uses, including drinking, fishing and washing, farm ponds have played an important role during the development of human civilization (Miller, 2009; Mushet et al., 2014; Chen et al., 2018). In recent decades, however, the development of reservoir-based water systems has gradually weakened the importance of farm ponds in water supply. Many areas that were once farm ponds have been filled as a result of expanding urbanization (Huang et al., 2012; Hill et al., 2018) and agricultural intensification (Calhoun et al., 2014; Poschlod and Braun-Reichert, 2017) or have become isolated from one another due to the modification of ditches and streams (Fang et al., 2014). Thus, some obvious questions arise: Do farm ponds have functions beyond those of irrigation and water supply? Should they be conserved from contemporary degradation, and what are the challenges and solutions?

Conservation theory and practice pertaining to small, scattered wetlands (e.g., vernal pools, prairie potholes, farm ponds, gilgais, and cypress domes) was developed in the 1980s (Rashford et al., 2011; Raebel et al., 2012; Calhoun et al., 2014; Hunter et al., 2017). Several countries have implemented national conservation programs, including the Vernal Pool Association in the US, the Millennium Ecosystem Assessment in France, and the Pond Action in England, in conjunction with transnational research collaborations, such as the European Pond Conservation Network (Reid et al., 2005; Biggs et al., 2010; Lassaletta et al., 2010). Along with numerous scientific papers, which have tripled in number over the last decade (source: Thomson Reuters' Web of Knowledge[SM], October 2018), several reviews have highlighted farm ponds as regional biodiversity hotspots, providing habitats and refuges for some of the most endangered wetland plants, invertebrates and amphibians (Céréghino et al., 2014; Mushet et al., 2014; Hunter et al., 2017; Hill et al., 2018). Although developing countries require more effort to combat the increasing crisis of freshwater resource scarcity, most programs and findings have focused on developed countries (Liu and Yang, 2012; UN-Water, 2018). According to FAO (2013), a shortage of freshwater resources in terms of both quantity and quality will occur in 1/5 of the developing countries by 2030 due to high agricultural water consumption and severe nonpoint source pollution. Moreover, the predominance of small farms makes the ponds more important to agricultural activities (e.g., paddy planting, fish farming, and unfenced poultry and livestock breeding) and rural life in peripheral areas (Mitsuo et al., 2014; Chen et al., 2018; Luo et al., 2018b). These circumstances point to a need for targeted and systematic analyses of farm ponds in the developing countries.

China is the largest developing country with a total population of 1.4 billion, with approximately 42% of the population living in rural areas. China has faced historical challenges in the conflict between irrigation water demand and spatiotemporally heterogeneous precipitation. To mitigate water shortages and regulate the distribution of water resources, farm ponds were first constructed in Anhui Province early in the 700s BC, and then with the southward movement of agriculture during approximately the next thousand years, their prevalence extended from the middle and lower reaches of the Yangtze River to the Pearl River Delta and Yunnan Province (Yin and Shan, 2001; Zhang et al., 2009). After the foundation of the People's Republic of China (PRC) in 1949, farm ponds were vigorously developed and flourished as a unique natural-economic-social complex across southern China (Verhoeven et al., 2006; Liu et al., 2009; Yu et al., 2015). In recent decades, however, farm ponds, particularly small ponds, have faced a severe decline in numbers and malfunctions in terms of agricultural and ecological services. Although several nationwide environmental protection policies and legislations have been enacted to remedy water quality deterioration and water ecology degradation (e.g., the Water Pollution Prevention and Control Action Plan issued by the Ministry of Environmental Protection in 2015, the Revised Water Pollution Prevention and Control Law issued by the National People's Congress in 2017, and the River/Lake Chief Mechanism issued by the General Office of the State Council in 2016), protection of farm ponds in upstream areas has been entirely ignored. Across China, there has been a lack of systematic effort to characterize, monitor, and assess the waters' environmental and ecological status (Fang et al., 2014; Yu et al., 2015). The lack of such efforts is mirrored in other developing countries in Eastern Europe, South Asia, and South America, where farm ponds are used extensively for agriculture, but are excluded in conservation programs for wetland ecosystems (Poschlod and Braun-Reichert, 2017; Hill et al., 2018).

In view of the above scientific challenges and policy deficiencies, this study aims to analyze the ecosystem services, degradation threats, and conservation strategies associated with farm ponds in southern China (Fig. 2.1). Although these water bodies exist in other regions across China, they are ubiquitous in southern China owing to the humid subtropical climate and prevalence of paddy planting. After a historical and spatial overview (Section 2), the importance of farm ponds over scales ranging from within-pond to regional is explored by reviewing historical documents and recent literature (Section 3). Current threats and management challenges for farm ponds are then investigated in the

Chinese context (Section 4), followed by a proposed conservation framework derived from, but not limited to, recognized policies and techniques used to manage similar wetland ecosystems (Section 5). This study provides the first synthetic perspective to assist policy-makers and researchers in understanding and restoring the environmental and ecological values of farm ponds in developing countries.

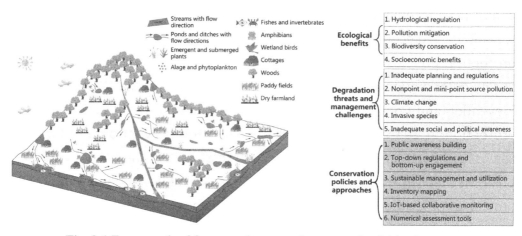

Fig. 2.1 Framework of farm pond perspective research of this chapter

2.2 Farm ponds in southern China

The history of farm ponds in China dates back to the Sinic civilization over 2,600 years ago. Quebei, the earliest pond on record, was constructed during the Spring and Autumn Period (770BC to 476BC) under the leadership of Sun Shu-Ao, who was the Prime Minister of the Chu State (Zhou et al., 2016a). As described in The *Commentary on the Waterways Classic*, Quebei received water from three streams and had five outlets for irrigation and flood discharge. Large-scale water conservancies became prevalent during the Han Dynasty (202 BC to 220 AD) as a result of the national policy of valuing agriculture above other industries. The total ponded area at that time was surprisingly large. For example, it was reported that the ponded area in Runan county, Henan Province, totaled 300 km^2, and its water storage capacity was far greater than that in the county in the 1970s (Hang, 2001). Owing to political infighting and warfare, the number of large ponds declined in both Tang and Song Dynasties (618 AD to 1279 AD). Even so, there was an increase in the number of small ponds, especially in the low mountainous and hilly areas of the Yangtze and Huai River basins. In the Qing Dynasty (1616 AD to 1911 AD), although pond irrigation was rarely recorded in historical files, small-scale water conservancies were widely implemented and voluntarily maintained by farmers (Lü and Chen, 2014). Along with the rapid population growth and an increasing food demand after the foundation of the PRC, small ponds flourished everywhere in southern China. However, their numbers and water storage capacity have declined sharply in recent decades due to flood damage, sludge blockage, and intentional filling or other destructive activities. As shown in Fig. 2.2, farm ponds played an indispensable role in China's traditional agriculture, and their number and function have been closely related to the policies of the central and local governments as well as the maintenance by farmers.

Fig. 2.2 Variations in the total number of farm ponds throughout Chinese history. A star indicates the amount estimated from references for a historical period highlighted in the same color. Additional comments are provided to explain some major features related to the farm ponds for each period. Citations: [a] Zhang (2009), [b] Li and Liu (1986), [c] Lü and Chen (2014), [d] Tan (2005), and [e] Hang (2001) and Yu et al. (2015).

Farm ponds refer specifically to the abovementioned small-scale water conservancies to distinguish them from artificial lakes and reservoirs (Chou et al., 2013; Gao et al., 2016). Although farm ponds are rarely mapped at large scales (Tiner et al., 2015; Wu, 2018), the latest Google imagery (source: the 2018 Digital Globe) offers a feasible means to view their shapes and distributions in southern China. As demonstrated by the eleven examples presented in Fig. 2.3, farm ponds are distributed across (1) almost every province and municipality, (2) four rainfall isohyets with average annual rainfall of 800 mm, 1,200 mm, 1,600 mm, and 2,000 mm, (3) landscapes with elevation ranging from 30 m to > 1,500 m, and (4) all four large river basins (the Yangtze River basin, the Pearl River basin, the southeastern river basins, and the southwestern river basins) and both island provinces (Taiwan and Hainan).

Fig. 2.3 Boundaries of southern China and examples of farm ponds. From left to right and top to bottom, the images show (1) Taoyuan county, Taiwan, (2) Youbu town, Zhejiang Province, (3) Dingan county, Hainan Province, (4) Hongshan town, Hunan Province, (5) Luchuan county, Guangxi Province, (6) Chenji town, Jiangsu Province, (7) Huanglu town, Anhui Province, (8) Huwei town, Chongqin Municipality, (9) Luliang county, Yunnan Province, (10) Yongping county, Yunnan Province, and (11) Hangwang town, Shanxi Province. An average elevation of the area is provided on the bottom right corner of each image.

Farm ponds have a variety of common names in Chinese, such as 堰 (yàn in Mandarin, literally meaning weir), 坝 (accordingly, bà and dyke), 陂 (bēi and pond), 堨 (ài and blockage), 捺 (nà and restraint), 塘 (táng and pond), and 荡 (dàng and marsh). These names were derived from local customs and traditions, and they are also used to reflect various topographic features of the surrounding landscape (e.g., plains, valleys, and hills), hydrologic and vegetation conditions, and embankment structures of a farm pond (Yu, 2015). Despite these differences, advances in measuring systems have enabled the quantification of pond characteristics in various dimensions. Specifically, a single pond can be described by its shape, area, perimeter, depth, hydroperiod (i.e., annual water level variation and flood frequency), and other parameters (Tan, 2005). From a watershed or landscape perspective, groups of ponds can be described by their density (i.e., area proportion of water surface to farmland), connectivity (i.e., whether there is water transport between ponds via ditches or streams), fragmentation (i.e., a measure of dispersion and aggregation), contagion (i.e., a measure of spatial distribution), and other indicators (Huang et al., 2012; Fang et al., 2014).

2.3 Why are these small and scattered waters important?

Farm ponds in conjunction with the surrounding woods, farmlands, cottages, and water courses, constitute a unique pastoral landscape of small farms in southern China (Fig. 2.4). While storing rainwater, farm ponds can profoundly alter the hydrologic and nutrient cycling in the watershed, and, as a consequence, lead to idiosyncratic riparian conditions to host flora and fauna. They also offer considerable social and cultural benefits, including improving physical and mental well-being and increasing awareness of environmental conservation at larger scales. The benefits of farm ponds are summarized in Table 2.1 and discussed in the sub-sections below.

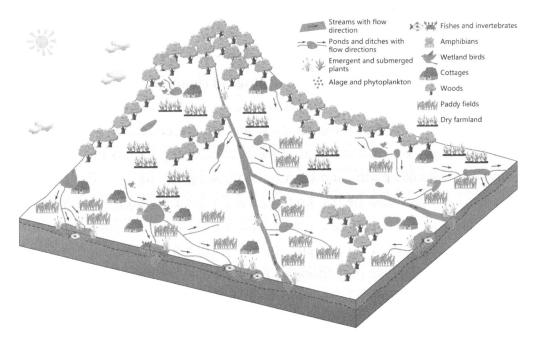

Fig. 2.4 A typical landscape with small, scattered farm ponds from low mountains to plains.

Table 2.1 Benefits of farm ponds in southern China

Categories	Functions
Hydrological regulation	(1) Intercept surface runoff, (2) store water for irrigation and rural life, (3) reduce and postpone flood peaks, (4) recharge and discharge groundwater, and (5) increase humidity via water surface evaporation.
Pollution mitigation	(1) Reduce nutrients from agricultural drainage, (2) intercept sediment in surface runoff, and (3) sequester atmospheric carbon dioxide.
Biodiversity conservation	(1) Provide nutrient source for the food web, (2) wildlife habitat, and particularly, (3) spawning and incubation environment for endangered species.
Socioeconomic benefits	(1) Mitigate flood risk, (2) increase farmers' income, (3) constitute pleasant scenery and improve mental well-being, and (4) provide educational and recreational opportunities for the broader community.

2.3.1 Hydrological regulation and ecological cleanup

Farm ponds provide disproportionate contributions to hydrological and ecological functions that have lag, sink, and source effects on the physical, chemical, and biological status of downstream waters (Rains et al., 2016). Rainfall initially fills ponds and then runs off along the surface. This process helps to reduce flood peaks, contributes to groundwater recharge and discharge, and provides stream base flow for irrigation. In addition to the qualitative descriptions in ancient literature, such as in *The Odes of Chen, Classic of Poetry*, on-site monitoring and numerical assessment have quantified those hydrological functions. For example, in the Liuchahe watershed, Anhui Province, the 193 farm ponds that accounted for 6.2% of the total watershed area were observed to reduce irrigation water shortages from 333 mm to 138 mm in wet years and from 280 mm to 41 mm in dry years. During a heavy rainfall storm event (141 mm), the ponds reduced the flood peak from 2.5 to 0.3 m^3/s (Yin and Shan, 2001). Pond parameters, such as shape, area, and depth, determine their water retention in the catchment, while their spatial distribution is important to the water storage capacity and utilization of rainwater in the watershed (Verhoeven et al., 2006; Fang et al., 2014).

Farm ponds can reduce diffuse pollutants from agricultural drainage via many processes including interception and precipitation, plant uptake, microbial degradation, and sediment adsorption. Capps et al. (2014) discovered that oxygen transport through the stems of emergent and floating plants can create an aerobic environment at their root surface, where organic matters can be decomposed by aerobic microorganisms and ammonia can be nitrified by nitrifying bacteria. In the low-oxygen areas of the deep root zone, where anaerobic fermentation is prevalent, organic matters can be denitrified into carbon dioxide and methane. Moreover, farm ponds can remove the particle-bound phosphorus in runoff through sedimentation and enhance the binding of soluble inorganic phosphate to Al^{3+} and Fe^{3+} in the sediments (Liu et al., 2009). A two-year monitoring in the irrigated area of Gaoyou county, Jiangsu Province, revealed that the concentration of total nitrogen and total phosphorus were reduced by 64.5 and 54.4%, respectively, by the groups of ponds and ditches (Peng et al., 2013). The reduction rate is determined by their connectivity, vegetation coverage and dredging operations (Mitsuo et al., 2014; Hansen and Dolph, 2018; Chen et al., 2018).

2.3.2 Biodiversity conservation

As an essential component of the circular "crop-livestock-fish" production system, farm ponds increase biodiversity from within-pond to regional scale. While crops are used to feed poultry and livestock, a large proportion of the manure produced by the poultry and livestock is applied into the ponds to promote fish growth, and the sediments in the ponds are returned to the land to improve soil fertility for crop production (Ongley et al., 2010; Chen et al., 2016a). Although most of the manure is consumed by fish, it provides large amounts of nutrients for accelerating the phytoplankton and zooplankton growth and supplies substrates for decomposing microorganisms. These increased biomasses, in turn, serve as food for a wide range of wildlife. In the ponds of the Liuchahe watershed, for example, Han and Sun (2007) reported a total of eight types of aquatic vegetation, including emergent plants, floating-leaved plants, and submerged plants. The vegetation covered approximately 12.6% of the pond surface area and formed different communities from the village, the farmland and the hillslope nearby. In the Taoyuan tableland, Huang et al. (2012) detected 94 bird species in pond areas. Serving as stepping stones between larger freshwater habitats, these ponds provide both a stopover site during winter migration for shorebirds and seabirds and attract many wetland birds to feed and nest. In addition, farm ponds are reported to provide habitats for many amphibians, such as frogs and toads, tortoises, and insects, including dragonflies, caddisflies, and midges. This even includes some species on the International Union for Conservation of Nature's Red List, which rely on the humid environment and abundant carbon resources in riparian areas, especially during the spawning and incubation periods (Raebel et al., 2012; Capps et al., 2014).

2.3.3 Socioeconomic benefits

A wealth of socioeconomic benefits stem from farm ponds. As enumerated above, the hydrological regulation and flood alleviation services contribute to social stability, and the improved nutrient cycling, particularly in the "crop-livestock-fish" system, is beneficial to agricultural production and farm economy. In a three-year study on abandoned ponds near Wuhan city, Hubei Province, for example, it was found that the annual fish production amounted 2.2 to 2.6 kg/m^2. This generated a profit of $1.96 to $2.55 (13.5 to 17.6 RMB, $1 ≈ 6.90 RMB) per square meter (Zhou et al., 2016b), and the whole city was estimated to benefit approximately $145 million (1 billion RMB) from the industry of pond aquaculture. Besides, *Phragmites* at the water's edge are reaped as raw material for papermaking in Hunan Province (Tian et al., 2011). Combined with farmlands, woods, and streams, farm ponds constitute a harmonious and nature-based rural residential scenery. For thousands of years, they have served as a traditional exchange place for neighbors' communicating and sharing life stories while washing, swimming and fishing. Such a water-centric lifestyle remains in the present, despite environmental and water quality deterioration due to nutrient enrichment from fertilizers and pesticides. In addition, these aquatic features may provide educational and recreational opportunities nationally and even globally. For example, Quebei has been honored as China's Key Relic Protection Unit since 1988 and a World's Heritage Irrigation Structure since 2015, attracting more than 100,000 domestic and foreign tourists annually (Zhou et al., 2016a). Although farm ponds triggered conflicts between different ethnic groups in northern Taiwan in the 1900s, a recent TV series called Beitang has aroused an appreciation of these land and water resources and introspection on their degradation in the Chinese community.

2.4 What are the threats and management challenges?

The most overarching and pervasive threat to farm ponds is linked to their small size and wide distribution. Anthropogenic disturbances and natural processes are the cause of farm pond degradation, while management and conservation need to involve policy, numerical assessment, public awareness, and cost-efficiency analysis (Table 2.2).

Table 2.2 Degradation threats and management challenges associated with farm ponds

Categories	Threats and challenges
Inadequate planning and regulations	(1) In ancient China, a government emphasis on agriculture determined the increase and decline in the number of farm ponds, (2) currently, urbanization, land consolidation, and centralized irrigation contribute to their degradation, and (3) environmental regulations ignore these water bodies.
Rural nonpoint and mini-point source pollution	(1) Nonpoint source pollution can damage ecosystems, cause an imbalance in material and energy flow, and degrade ecosystem services, (2) mini-point source pollution from animal feeding operations and rural households is also significant, and (3) ponds are less capable of diluting pollutants.

Continued

Categories	Threats and challenges
Climate change	(1) Climate change can lead to increased air temperature, rising sea levels, and more frequent climatic extremes, and can also (2) cause abnormal hydroperiods and negatively affect invertebrate diversity and plant species composition.
Invasive species	(1) Invasive species, including plants, crayfish, and amphibians, (2) can be harmful to the habitat structure and human health, and (3) may lead to secondary pollution if biological and chemical treatment are implemented.
Management	(1) Requires consideration of material, organisms and energy fluxes at various spatial scales, (2) needs to involve awareness and understanding of the ecosystem services in the broader community, and (3) should be better quantified regarding the ecosystem services and conservation costs.

2.4.1 Inadequate planning and regulations

The lack of rigor and consistency in planning small water conservancies is a widespread and chronic phenomenon (McGreavy et al., 2012; Poschlod and Braun-Reichert, 2017; UN-Water, 2018). Since ancient times, the construction and maintenance of farm ponds has been driven by the government prioritization of agriculture rather than assessment of the abovementioned ecosystem services. For example, in the golden age of the Han Dynasty, pond projects were launched nationwide, and more than 60% of the ponds were located in the agriculturally prosperous Yangtze and Huai River basins (Zhang, 2009). When it came to the subsequent Three Kingdoms Period, however, the ponds were damaged to restrict agriculture in neighboring realms, with Quebei losing 60% of its water surface area due to warfare (Hang, 2001). Today, China is facing grand challenges in feeding 22% of the world's population and land uses have been largely intensified to maintain the current average annual growth rate in agricultural production to be as high as 4.6%. As a result, many farm ponds have been converted to paddy fields, orchards, or threshing grounds (Yin and Shan, 2001). Moreover, centralized irrigation based on reservoirs and canals has weakened the role of farm ponds in water supply, resulting in further negligence and mismanagement of the ponds and increased vulnerability to floods and habitat losses (Liu and Yang, 2012).

Although recent evidence indicates that farm ponds contribute disproportionately more to aquatic biodiversity than larger and more widely studied freshwater systems, such as lakes and rivers, environmental legislation and management strategies of these features is far behind the needs due to the ignorance by the government. In Europe, the EU Water Framework Directive (WFD; 2000/60/EC) was implemented to protect and improve the quality of ground and surface waters, but, in practice, it only covers rivers and standing waters with a surface water area larger than 0.5 km2, therefore excluding the vast majority of temporary streams and wetlands (Lassaletta et al., 2010). Similarly, legislation in the US has protected the nation's navigable waters under the Clean Water Act since 1972, but these rules did not consider isolated and non-navigable wetlands as a "significant nexus" within the hydrological system before two heated debates in 2001 and 2006 (Leibowitz et al., 2008). China has recently spent millions of dollars (billions of RMB) to remedy widespread environmental and ecological deterioration through various actions including converting farmlands to forests and grasslands, establishing wastewater treatment plants, and removing algae in lakes and rivers (Huang et al., 2019). Although these programs, in conjunction with management strategies, such as the River/Lake Chief Mechanism, and examination criteria, such as the Environmental Quality Standards for Surface Water (GB 3838-2002), have obtained phased achievements, the protection of farm ponds in upstream areas has been entirely ignored.

2.4.2 Rural nonpoint and mini-point source pollution

The extensive use of synthetic fertilizers and manure is a major characteristic of modern agriculture, and the associated discharge of nitrogen and phosphorus has become a major source of nonpoint pollution in rural areas. These pollutants can damage ecosystems, impair material and energy flows, and reduce or eliminate ecosystem services. For example, ammonia can increase the pH and temperature of ponded waters and can be toxic to fish and invertebrates (Zheng et al., 2017). At elevated concentrations, nitrogen and phosphorus are the main causes of algal blooms in surface water, while nitrate in drinking water sources is linked to birth defects, methemoglobinemia in infants, and high blood pressure in adults (Ongley et al., 2010; Chen et al., 2016a). In addition, it is noteworthy that farm ponds are more vulnerable to pollution than larger water bodies, due to their smaller capacity of diluting pollutants physically, chemically, or biologically, especially when the surrounding drainage network is modified (Huang et al., 2012; Céréghino et al., 2014).

Although intensive point source discharge, such as industrial wastewater, is rare in agricultural regions, mini-point pollution sources, including animal feeding operations and domestic sewage and wastes, pose nonnegligible threats to the ecosystems of farm ponds. According to the First National Census of Pollution Sources (Ministry of Agriculture of the PRC, 2009), poultry and livestock feeding produces 12.7 Tg (1 Tg = 10^{12} g) total nitrogen and 0.2 Tg total phosphorus annually, of which approximately 43.6% originates from unfenced feeding operations. Scattered domestic animals live in riparian areas, discharging their manure and urine into the water. Domestic sewage in these areas is barely treated before being discharged into ponds and ditches owing to the lack of sewage collection and treatment facilities (Ongley et al., 2010). In addition, domestic wastes, including plastic bags, obsolete electrical appliances, and food residues are usually discarded into the water, resulting in persistent microplastic and heavy metal pollution as well as unpleasant colors and smells (Zhang et al., 2018).

2.4.3 Climate change

Farm ponds, owing to their small catchment area and size, are vulnerable to changes in temperature and precipitation patterns. According to the Fifth

Assessment Report of the Intergovernmental Panel on Climate Change (IPCC), the increase in temperature was the highest during the past century, while extreme precipitation and storm events have increased dramatically, and sea levels have risen 0.17 to 0.21 m in the Northern Hemisphere (Stocker et al., 2014). In addition, owing to the combined effects of El Niño and La Niña events, East Asia, especially the coastal provinces of southeastern China, usually suffers continuous precipitation and floods from early spring through the entire monsoon period (i.e., March to July, Karori et al., 2013; Luo et al., 2018a). These climatic extremes can cause an abnormal hydroperiod and negatively affect invertebrate diversity and plant species composition. For example, widespread droughts occurred almost every year in the southwestern regions during the past two decades, and many ponds dried up and were abandoned (Han et al., 2014). In the southeastern regions, however, soil erosion tends to be severe in the low mountains and hills, where frequent dredging operations are required to avoid siltation, eutrophication, and flooding.

2.4.4 Invasive species

Invasive species can pose a serious threat to freshwater biodiversity. China's humid weather conditions, widespread transportation network, and fragmented habitat of native species have provided favorable conditions for exotic and invasive species. Invasive species in farm ponds include plants, crayfishes, and amphibians (Yan et al., 2001; Wang et al., 2017). For example, the growth and propagation of water hyacinth can cover the entire water surface, blocking light, suppressing the growth of plankton, and causing severe degradation of other aquatic flora and fauna. These plants can even be harmful to human health, promoting the breeding of rats, flies, and other carriers of bacteria, if the waters simultaneously contain domestic wastes. To reduce such adverse effects, the Shanghai government has spent $2.18 million (15 million RMB) annually to remove water hyacinth in the suburbs, but some biological and chemical agents have resulted in secondary pollution (Lu et al., 2007).

The above natural and anthropogenic threats can be significant by their own, but the interactions of the changing climate, invasive species, and landscape degradation due to increasing demands for food, space, and water resources make the need for suitable management strategies clear if we are to sustain the multifarious benefits of these aquatic features.

2.4.5 Management challenges

Farm ponds act as biological, physicochemical, and ecological hotspots within an agricultural matrix and require management at both the within-pond and landscape scales (Chou et al., 2013; Mushet et al., 2014; Hill et al., 2018). Because the hydroperiod of farm ponds is mainly determined by the local climate, microtopography, and water consumption, they are extremely susceptible to changes in the surrounding land uses and crop types. Due to their small water area and volume, farm ponds are also very sensitive to alterations in chemistry of sediments and pollutants. In addition, farm ponds usually support life forms of amphibians from larva to adult and migratory birds from hundreds of miles away, making the adjacent terrestrial habitat an integral part of their ecosystem services. Direct losses or degeneration of these water bodies decreases the regional wetland density and increases travel distances for wildlife, particularly those relying on multiple aquatic resources. In the past, studies on farm ponds have been focused on their single ecosystem service (Chou et al., 2013; Gao et al., 2016). Integrated pond management is usually difficult to achieve due to predefined priorities, funding limitations, and narrow research objectives that lack the scientific understanding of those multi-scale processes (Spiteri and Nepalz, 2006; Raebel et al., 2012). Therefore, one major challenge is to manage, conserve, and restore ponds by integrating several spatial scales, while taking into account all fluxes of materials, organisms, and energy. Although it is very difficult, being aware of the implications of pond management across these aspects is important.

Limited public awareness and understanding of small, scattered wetlands complicates the management of farm ponds. Compared with agricultural irrigation and flood alleviation, the less visible ecosystem services (e.g., providing habitats for endangered species and contributing to groundwater recharge and discharge) are not widely recognized, which diminishes local support for public conservation actions (Hunter, 2005; Zhang, 2009; Golden et al., 2014). On the other hand, some distinguished natural landscapes usually have a flagship wildlife species to capture the hearts and minds of the public, such as pandas for the Chinese bamboo forests and penguins for the Antarctic ice sheets. Although migratory birds or amphibians may be good candidates if marketed well, farm ponds lack such charismatic species but can be regarded as breeding havens for some unpleasing animals, such as poisonous snakes, mosquitoes and leeches. Another challenge is therefore to improve the broader communities' understanding of the ecosystem services even though farm ponds seem only related to the rural households nearby.

The small size, large number, and wide distribution of farm ponds make developing conservation strategies confusing and costly. Ecosystem services of farm ponds must be better articulated to change the perceptions of the public and authorities. Some naturalist associations and scientific organizations, such as the European Pond Conservation Network, the Million Ponds Project in the UK, and the Taoyuan Pond and Waterway Conservation, are active in compiling documents, illustrations, and recommendations, but few provide numerical assessments to assist practical conservation actions. Widespread benefits and concentrated costs can challenge conservation strategies based on negotiations with individual landowners who have ponds on their property but envision a more profitable use of the land (Shogren et al., 2003). This problem is more difficult in China, as small water conservancies are formally state-owned, but paradoxically, the dredging and reinforcement of farm ponds are generally funded by local communities (Tan, 2005). While the ecological and social importance of farm ponds can be quite significant, only limited value is offered to small farmers who have ponds in their fields, which decreases their interest in conserving the projects. Therefore, management policies and approaches that quantify ecosystem services and recognize the full extent of conservation costs are more likely to succeed in navigating these challenges.

2.5 Policies and approaches for farm pond conservation

If farm ponds are conceptualized as complex ecosystems, their role in wetland conservation planning is quite evident. In this section, we address three conservation policies, including public awareness building, top-down regulations and bottom-up engagement, and sustainable management and utilization, as well as three technical approaches, including inventory mapping, IoT-based collaborative monitoring, and numerical assessment, from similar practices in, but not limited to, wetland protection. These methods are oriented to the small, scattered characteristics of farm ponds, and involve various stakeholders, including community members and government sectors, to ensure a long-term, collaborative conservation framework in developing countries. Fig. 2.5 shows how these methods, stakeholders and pond landscape may fit together logically, and each of the methods is discussed in the sub-sections below.

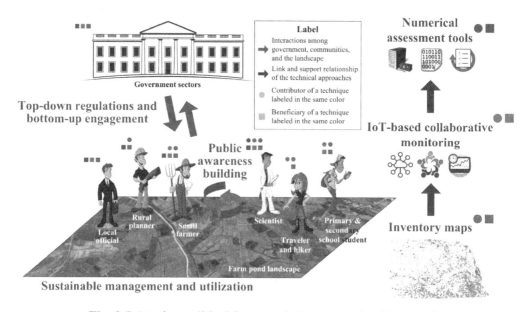

Fig. 2.5 A technopolitical framework for conserving farm ponds

2.5.1 Public awareness building

The critical first step of public awareness building is to ensure that the terms "small wetlands" "freshwater ecosystems" and "landscape degradation" are in the lexicon of farmers, local officials, rural planners, and urban visitors. Education serves as a formal and fundamental method to build such environmental awareness and to make conservation knowledge explicit and widespread (Varela-Losada et al., 2016). In addition to classroom education programs, on-site environmental activities can enhance student learning and instill an intrinsic valuation of nature. For example, survey data indicated that primary school students exhibited increased interest in soil and water sciences and willingness to protect red-crowned cranes after four seasonal visits to the Zhalong Nature Reserve, northeastern China (Zhao, 2006). Environmental-friendly attitudes and behaviors have also been demonstrated to be transferable from children to their families (Damerell et al., 2013), which is important because animal feeding operations and household water management are closely related to the ecosystems of farm ponds.

Informal education by means of newspapers, radio, television, and the internet, is more flexible and diverse, translating scientific findings into policy and converting environmental knowledge into action for various stakeholders (Greenhow and Lewi, 2016). For example, the Vernal Pool Association has operated a promotional website since 2005 and identified more than 300 vernal pools in conjunction with their invertebrates' lifecycles in Massachusetts, US, where the filling and draining of ephemeral ponded wetlands is prohibited. Similarly, in the intertidal zones of Zhejiang Province, 8,000 popular science readings and 1,000 wildlife photo albums were compiled and disseminated to rural residents during World Wetlands Day and Bird Loving Week from 2011 to 2014. These substantial investments in public education have helped to implement a strong regulatory basis for the restoration of mangrove swamp and estuarine wetland (Tang, 2015).

2.5.2 Top-down regulations and bottom-up engagement

Regulations promoting the conservation and management of small, scattered wetlands may range from restrictive local protections at the pond scale and across hundreds of meters of adjacent habitat at the landscape scale to broader national guidelines and international commitments (Biggs et al., 2010; Fang et al., 2014; Mitsuo et al., 2014). At any governmental level, top-down regulations have the advantage of setting clear rules but usually lack local understanding and enforcement jurisdiction of the various stakeholders being regulated. This is particularly true for farm ponds, which are sometimes closely related to agricultural activities and rural life. Hence, their management may best be achieved by combining top-down approaches, which set an overall standard, and bottom-up engagement, which ensures effective complements via more tailored and voluntary actions. For example, vernal pool conservation regulations at the federal and state levels only include formal definitions, criteria to support pool prioritization, and a basis for delimiting buffer zones, while local programs are more concerned with the economic benefits of land use and the surrounding flora and fauna in Maine, US (Calhoun et al., 2014).

Top-down regulations have gradually shifted from rivers and lakes to small, manmade, or natural shallow waters in developed countries, as the latter are numerous, typically outnumbering the larger waters, and can be regarded as biogeochemical hotspots in virtually all terrestrial environments (Leibowitz et al., 2018). For example, the US Environmental Protection Agency concluded that ponded wetlands are effective in riverine nitrate removal and thus are "waters of the US" under the Clean Water Act (Leibowitz et al., 2008). The EU Water Framework Directive has recently recognized the importance of intermittent headwater streams due to their vulnerability to anthropogenic pressures and significant influence on downstream water quality (Lassaletta et al., 2010). However, such "legal" standing for farm ponds is lacking in China's environmental legislation. Following the institutional reform of the State Council in 2018, the Ministry of Ecology and Environment, together with the Ministry of Natural Resources, Ministry of Water Resources, and Ministry of Agriculture and Rural Affairs, recommended standardization of construction, operation, and maintenance of farm ponds, especially those in disrepair or in areas vulnerable to floods and landslides. These ministries would also be responsible for the proposal of relevant laws and regulations, such as "Procedures to Build a Farm Pond" "Plans for Farm Pond Protection", and "Farm Pond Restoration Provisions".

Meanwhile, bottom-up initiatives must be encouraged. On-site pond conservation programs usually require understanding of natural conditions (e.g., hydrology, soil, and geology) and agricultural practices in peripheral areas. The involvement of local communities can provide such information and also help to enhance the effectiveness and continuous improvement of those programs (Calhoun et al., 2014). Citizen science has become a popular way to engage "nonprofessionals" in ecological assessment and environmental protection across scattered areas (Zheng et al., 2017). With regard to farm ponds, recruiting voluntary or somewhat incentive-based farmer scientists and holding regular workshops for their family members is recommended to foster a sense of responsibility, mutual surveillance, and eventually collaborative conservation, as has been tested in farmer-participatory rural reform in the Chenzhuang village, Jiangsu Province (Yang and Chen, 2017; Chen et al., 2018).

Like participant incentives, conserving natural resources is also sometimes expensive and time-consuming owing to the need for infrastructure, maintenance, informational material, etc. Therefore, it is essential to develop partnerships with different communities and identify opportunities to meet disparate goals with desirable outcomes for various stakeholders (Spiteri and Nepalz, 2006; McGreavy et al., 2012). In southern Saskatchewan, Canada, for example, prairie pothole protection is combined with govern funding and investments from nongovernmental organizations (Rashford et al., 2011). To introduce up-to-date technologies and management skills in urban stormwater management, Chinese authorities have adopted the participation of private and foreign investors since the late 1990s. Their cooperation has started a unique form of Public-Private Partnership (PPP) projects, which take advantage of top-down regulation from government sectors and the flexible operation of market-oriented entities (Lee, 2010). Despite potential operational risks, including postponed approvals and imperfect supervision, these diversified opportunities are referential to farm pond conservation, especially when funds and techniques are limited.

2.5.3 Sustainable management and utilization

Small, scattered wetlands are arguably best managed using the meso-filter strategy (Hunter, 2005; Hunter et al., 2017), where features that may be integrated ecosystems in their own right, or ecological elements within a larger context can, by nature of their small size, move towards sustainable management. As a unique natural-economic-social complex, farm ponds also require appropriate management to collect stormwater, mitigate flood risk, and provide considerable economic and cultural benefits in rural areas (Yin and Shan, 2001; Mitsuo et al., 2014; Yu et al., 2015). In view of their importance to both nature and humanity, such management needs to incorporate approaches that range from within-pond practices specific to water and sediments to landscape-scale strategies that recognize their functions as wetland complexes that are embedded within, and may be integral to, larger ecosystems.

From historical documents on paddy planting to recent water conservancy practices, farm pond management, including dam design and construction and pond repair and dredging, have always been highlighted in southern China. For example, early in the Spring and Autumn Period, *The Records of Examination of Craftsman* proposed that a dam's cross-section should be trapezoidal in shape and the slope of the trapezoid edge is recommended to be 1:1.5 to 1:2.0 for structural stability. With regard to regional pond planning, *The Huainanzi Shuolin Discourse* compiled in the Han Dynasty stated that waters with a total area of 0.1 km^2 and an average depth of 2m can mitigate droughts in surrounding farmland of 0.4 km^2, indicating that a pond density of

25% is appropriate for agricultural irrigation. Pond repair and dredging were biyearly carried out after crop harvest between the Yangtze and Huai River (Lü and Chen, 2014). Despite increasing flood risk and soil erosion due to climate change and a dwindling number of small farmers in the countryside, these management traditions are indispensable to maintain water capacity, improve water quality, and promote nutrient cycling.

Protecting farm ponds from degradation and destruction is a multi-scale issue that requires broader perspectives and strategies. For example, conservation prompts, such as "Don't litter" signs, have proven useful to reduce floating wastes from rural households and tourists, as they are highly specific, proximal to a site, and can be stated in a friendly manner to motivate one's responsibility (Osbaldiston and Schott, 2012). In addition, following the experience of water caltrop planting in Anhui Province and *Phragmites* growing in Hunan Province (Liu et al., 2009; Tian et al., 2011), special attention needs to be paid to aquatic vegetation, as it provides habitat to amphibians and benthic invertebrates and is easily consumed by the farmed fishes. In addition, low-impact development, including the use of pervious pavement and the construction of rain gardens and forested buffers in riparian areas, is encouraged in rural areas, as it has been included in urban planning. When constructed at an optimized density and connectivity, farm ponds can act as nature-based "sponges" (Yu et al., 2015; UN-Water, 2018), reducing nutrient loads into downstream waters (Hansen et al., 2018; Leibowitz et al., 2018).

2.5.4 Inventory mapping

To effectively manage and conserve farm ponds individually and at a landscape scale, it is essential to have a spatially explicit inventory combined with a collection of the hydrological and ecological status of the ponds, including information on the adjacent agricultural matrix. Remote sensing methods have been widely used to analyze pond dynamics, e.g., the loss rate and driving forces in the Taoyuan tableland (Fang et al., 2014), but farm ponds are rarely mapped at large scales, as they are usually disregarded or not documented in places where wetland inventories are compiled and are difficult to automatically update from high-resolution, nationwide imagery (Tiner et al., 2015; Wu et al., 2018). In addition, the ability to identify farm ponds varies greatly depending on the natural conditions, such as water level fluctuations and adjacent vegetation coverage. For example, in evergreen and mixed forest landscapes in the northeastern US, only 43% of the confirmed temporary pools were detected by aerial photography. Even in open areas, remote detection can be limited by atmospheric conditions and the spatial resolution of the sensors being used (DiBello et al., 2016). However, technological advances are constantly improving this work. The availability of high-resolution light detection and ranging (LiDAR), synthetic aperture radar (SAR), hyperspectral, and multispectral data has increased the feasibility of detecting water bodies as small as 1 1 m. Moreover, the introduction of multisensory and multi-scale data fusion techniques has enabled the large-scale mapping of small wetlands with unprecedented accuracy (Tiner et al., 2015).

Despite the abovementioned management constraints and technological challenges, inventory maps of ponded waters can be developed at multiple scales from field surveys, image interpretation, and voluntary programs using farmer scientists to identify their location, extent, and ecosystem structure (Calhoun et al., 2014), to conduct large-scale inventories associated with hydrological and ecological assessments to better support research, management, and government initiatives, as has been done in other specialized databases with numerous, scattered points of interest (Chen et al., 2016b).

2.5.5 IoT-based collaborative monitoring

Recent developments in environmental monitoring have enabled quantitative descriptions of the physical, chemical, and biological characteristics of water bodies over time and space. In this process, one of the most important features is the Internet of Things (IoT), which provides a network of various smart devices that sense, interpret, and react to the environment (Perumal et al., 2015). Owing to their low cost, ease of access, and tailored devices, IoT-based techniques have inspired engineers and private entities to either build their own sensors or modify off-the-shelf equipment to realize the integrated and real-time detection of hydrological, water quality, aquatic biota and soil conditions. For example, in Hubei, Jiangsu, Guangdong, and other provinces, connected sensors coupled with imbedded knowledge base have facilitated the aquatic plant growth, nutrient management, and disease prevention in shrimp breeding ponds, transforming traditional inland fisheries and aquaculture into data-driven and intelligent activities (Ma et al., 2012).

Collaborative efforts from different communities can expand the horizons of environmental monitoring and complement infrastructure and fixed observation stations. For example, mobile phones with GPS, photography, and social media applications have proven effective in collecting data on the color and transparency of surface waters by citizens, as has been implemented in the River Chief Mechanism (Zheng et al., 2017). Simple and inexpensive water test kits, available from many pet stores or online retailers, are useful to small farmers when preventing water quality degradation in fish ponds (Naigaga et al., 2017). High-throughput sequencing of DNA extracted from water samples serves as a quick, cost-effective and standardized method for scientists and conservation agencies when evaluating rare and threatened species across a wide range of taxonomic groups (Thomsen et al., 2012). By properly integrating these discrete results with consecutive data, IoT-based monitoring can be extended to a human-centered network with wide community involvement. Fig. 2.6 shows an integrated monitoring system being developed for farm pond conservation in the Chenzhuang village. Although this is a long-term effort, the GIS-based data framework will provide a solid basis for understanding pond ecosystem services.

2.5.6 Numerical assessment tools

To balance the costs and benefits of farm pond conservation, the hydrological, biogeochemical, and biodiversity functions that support ecosystem services and socioeconomic value should be assessed in a standalone or

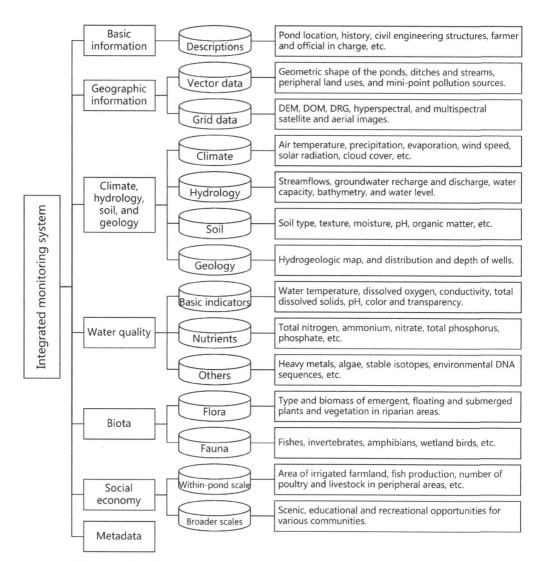

Fig. 2.6 The data framework of an integrated monitoring system for farm pond conservation

comprehensive manner (Fang et al., 2014; Hill et al., 2018). For example, based on streamflow observations, Gao et al. (2016) discovered a decreasing trend of water retention from downstream to upstream in the pond landscape of Chenji town. Chou et al. (2013) employed questionnaires, the fuzzy Delphi method, and an analytic hierarchy to assess the habitat suitability of pond-breeding amphibians in Yunlin County, Taiwan, China. Taking aquatic environments, terrestrial settings, and landscape connectivity into account, their results revealed that 19% of the 481 farm ponds were rich in frog species with high conservation value. In contrast to these single-criteria analyses, the vernal pool assessment in the northeastern US relies on both indoor and outdoor citizen-training sessions, as the former requires aerial photointerpretation and GIS to delineate waters, and the latter includes field surveys of egg masses, amphibians, and invertebrates (McGreavy et al., 2012; DiBello et al., 2016). Applied in a larger context, this method can provide a detailed assessment of aquatic features but requires labor-intensive efforts to establish a sound monitoring database beforehand.

Process-based models, which are based on a theoretical understanding of ecological processes and a set of computer programs, are useful to assess wetland dynamics and associated conservation decisions, especially when the required data are discrete and deficient (Ongley et al., 2010). Popular model categories include (1) watershed models, such as SWAT, HSPF and AnnAGNPS, which describe rainfall-runoff processes and often include an associated biogeochemical module that routes point and nonpoint source pollutants from the landscape to the stream, (2) hydrodynamic and water quality models, such as WASP, FEWMS and EFDC, which emphasize flow circulation, pollutant fate and transport and the interactions among the sediment, nutrients, phytoplankton and macrophytes in surface waters, (3) groundwater models, such as MODFLOW and HydroGeoSphere, which estimate the movement of subsurface flows through saturated porous media, and (4) ecosystem models, such as AquaSim, Ecopath with Ecosim and AEMON, which quantify the ecological relationships and evaluate the costs and benefits of wildlife conservation (Golden et al., 2014; Janssen et al., 2015; Rains et al., 2016). Although these models are deliberate in their respective domains, detailed modeling of pond dynamics is more complex, as it simultaneously includes surface and subsurface hydrological processes, biogeochemical cycles in ponds and ditches, and ecological processes at various spatial scales (Rains et al.,

2016; Chen et al., 2018). Therefore, appropriate integration of these models is expected to assess the ecosystem services of farm ponds individually and comprehensively.

To make the modeling tools more immersive and comprehensible, several emerging techniques can be employed, such as serious games and gamification and augmented and virtual reality, which have been tested in land management (Schulze et al., 2015), water resource protection for over-pumped aquifers, and climate change adaption involving sociopolitical events, energy consumption, population growth, etc. These technical advances are helpful to illustrate interdisciplinary teaching and education regarding pond conservation and the complex interrelations between the ponds and human well-being.

2.6 Conclusions

Nature-based solutions are becoming increasingly recognized as important for addressing the complex challenges in hydrology, ecology, and biodiversity. The multiple functions of small, scattered wetlands that support ecosystem services and their socioeconomic benefits in different countries make these water bodies interesting to scientists and decision-makers. As unique pastoral landscapes in rural areas, farm ponds have represented such aquatic features throughout history in southern China but have not received the attention they deserve, especially in terms of government regulation and environmental policies. Urbanization in conjunction with land consolidation and the prevailing reservoir-based irrigation systems have now engendered the degradation of these small wetlands, while rural nonpoint and mini-point source pollution, increasing climatic extremes, and invasive species have expedited water quality deterioration and adjacent ecosystem degradation.

A wide array of conservation strategies is proposed, starting with public awareness building and inventory mapping, thus creating a foundation for appropriate conservation. Top-down regulations and bottom-up engagement are simultaneously encouraged to ensure policy effectiveness and public initiatives, while sustainable management and utilization are necessary to reduce adverse influences on the aquatic ecosystems. Ultimately, with wide participation, IoT-based collaborative monitoring and the integration of process-based models are recommended to quantify the ecological processes, ecosystem services, and cost-efficiency of conservation programs. As a first synthetic perspective to emphasize the ecological roles of farm ponds in agriculturally dominated developing countries, this paper calls for attention to these endangered water bodies and encourages sustainable rural development among policy-makers, stakeholders, and the general public. The analytical framework and comprehensive suggestions are also referential to other small, scattered wetlands for which restoration efforts are required.

References

Biggs, J., Walker, D., Whitfield, M., et al., 2010. "Pond action: Promoting the conservation of ponds in Britain." *Freshwater Forum* 1: 2.

Calhoun, A.J., Jansujwicz, J.S., Bell, K.P., et al., 2014. "Improving management of small natural features on private lands by negotiating the science-policy boundary for Maine vernal pools." *Proceedings of the National Academy of Sciences* 111(30): 11002–11006.

Capps, K.A., Rancatti, R., Tomczyk, N., et al., 2014. "Biogeochemical hotspots in forested landscapes: The role of vernal pools in denitrification and organic matter processing." *Ecosystems* 17(8): 1455-1468.

Céréghino, R., Boix, D., Cauchie, H.M., et al., 2014. "The ecological role of ponds in a changing world." *Hydrobiologia* 723(1): 1-6.

Chen, M., Sun, F., Shindo, J., 2016a. "China's agricultural nitrogen flows in 2011: Environmental assessment and management scenarios." *Resources, Conservation and Recycling* 111(4): 10-27.

Chen, W., He, B., Nover, D., et al., 2018. "Spatiotemporal patterns and source attribution of nitrogen pollution in a typical headwater agricultural watershed in Southeastern China." *Environmental Science and Pollution Research* 25(3): 2756-2773.

Chen, W., He, B., Zhang, L., et al., 2016b. "Developing an integrated 2D and 3D WebGIS-based platform for effective landslide hazard management." *International Journal of Disaster Risk Reduction* 20: 26-38.

Chou, W.W., Lee, S.H., Wu, C.F., 2013. "Evaluation of the preservation value and location of farm ponds in Yunlin County, Taiwan." *International Journal of Environmental Research and Public Health* 11(1): 548-572.

Damerell, P., Howe, C., Milner-Gulland, E.J., 2013. "Child-orientated environmental education influences adult knowledge and household behaviour." *Environmental Research Letters* 8(1): 015016.

DiBello, F.J., Calhoun, A.J., Morgan, D.E., et al., 2016. "Efficiency and detection accuracy using print and digital stereo aerial photography for remotely mapping vernal pools in New England landscapes." *Wetlands.* 36(3): 505-514.

Fang, W.T., Chou, J.Y., Lu, S.Y., 2014. "Simple patchy-based simulators used to explore pondscape systematic dynamics." *PloS One* 9(1): e86888.

FAO, 2013. FAO Statistical Yearbook: World Food and Agriculture. Food and Agriculture Organization of the United Nations, Rome.

Gao, P.F., Li, Y.F., Liu, H.Y., et al., 2016. "Evaluation the contribution capacity of water conservation of pond system: A case study of Fengling Watershed." *Geography and Geo-Information Sciences* 6: 94-100.

Golden, H.E., Lane, C.R., Amatya, D.M., et al., 2014. "Hydrologic connectivity between geographically isolated wetlands and surface water systems: A review of select modeling methods." *Environmental Modelling & Software* 53: 190-206.

Greenhow, C., and Lewin, C., 2016. "Social media and education: Reconceptualizing the boundaries of formal and informal learning." *learning media & technology* 41(1): 6-30.

Grobicki, A., Chalmers, C., Jennings, E., et al., 2016. "An Introduction to the Ramsar Convention on Wetlands (previously The Ramsar Convention Manual)." Ramsar Convention Secretariat, Gland, Switzerland.

Han, L.Y., Zhang, Q., Yao, Y.B., et al., 2014. "Characteristics and origins of drought disasters in Southwest China in nearly 60 years." *Acta Geographica Sinica* 69(5): 632-639.

Han, X.Y., and Sun, P., 2007. "Aquatic macrophyte vegetation restoration in multi-pond wetland systems and its control on nonpoint source pollution." *Journal of Hefei University* 17: 71–74.

Hang, H.Q., 2001. "Positive and negative effect of Beitang's ups and downs and its thinking." *Ancient and Modern Agriculture* 4: 12-17.

Hansen, A.T., Dolph, C.L., Foufoula-Georgiou, E., et al., 2018. "Contribution of wetlands to nitrate removal at the watershed scale." *Nature Geoscience* 11(2): 127-132.

Hill, M.J., Hassall, C., Oertli, B., et al., 2018. "New policy directions for global pond conservation." *Conservation Letters* 11(5): e12447.

Huang, J., Zhang, Y., Arhonditsis, G.B., et al., 2019. "How successful are the restoration efforts of China's lakes and reservoirs?" *Environment International* 123: 96-103.

Huang, S.L., Lee, Y.C., Budd, W.W., et al., 2012. "Analysis of changes in farm pond network connectivity in the peri-urban landscape of the Taoyuan area Taiwan." *Journal of Environmental Management* 49(4): 915-928.

Hunter, J.M.L., 2005. "A mesofilter conservation strategy to complement fine and coarse filters." *Conservation Biology* 19(4): 1025-1029.

Hunter, J.M.L., Acuña, V., Bauer, D.M., et al., 2017. "Conserving small natural features with large ecological roles: A synthetic overview." *Biological Conservation* 211: 88-95.

Janssen, A.B., Arhonditsis, G.B., Beusen, A., et al., 2015. "Exploring, exploiting and evolving diversity of aquatic ecosystem models: A community perspective." *Aquat. Ecol.* 49(4): 513-548.

Karori, M.A., Li, J., Jin, F.F., 2013. "The asymmetric influence of the two types of El Niño and La Niña on summer rainfall over Southeast China." *Journal of Climate* 26(13): 4567-4582.

Lassaletta, L., García-Gómez, H., Gimeno, B.S., et al., 2010. "Headwater streams: Neglected ecosystems in the EU Water Framework Directive. Implications for nitrogen pollution control. Environ." *Environmental Science & Policy* 13(5): 423-433.

Lee, S., 2010. "Development of public private partnership (PPP) projects in the Chinese water sector." *Water Resources Management* 24(9): 1925-1945.

Leibowitz, S.G., Wigington, J.P. J., Schofield, K.A., et al., 2018. "Connectivity of Streams and Wetlands to Downstream Waters: An Integrated Systems Framework." *Journal of the American Water Resources Association* 54(2): 298-322.

Leibowitz, S.G., Wigington, P.J., Rains, M.C., et al., 2008. "Non-navigable streams and adjacent wetlands: Addressing science needs following the Supreme Court's Rapanos decision." *Frontiers in Ecology & the Environment* 6(7): 364-371.

Li, X.C., and Liu, X., 1986. "Abandonment causes of Beitang in ancient China." *Agricultural History of China* 3: 26-37.

Liu, J., and Yang, W., "2012. Water sustainability for China and beyond." *Science* 337(6095): 649-650.

Liu, Y., Fu, Q., Yin, C., 2009. "Phosphorus sorption and sedimentation in a multi-pond system within a headstream agricultural watershed." *Water Quality Research Journal of Canada* 44(3): 243-252.

Lu, J., Wu, J., Fu, Z., et al., 2007. "Water hyacinth in China: A sustainability science-based management framework." *Environ. Manage.* 40(6): 823.

Lü, J.L., and Chen, E.H., 2014. "Discussion on water resources environment and the construction of dam in Chaohu lake basin in Ming and Qing dynasties." *Historiography Research in Anhu* 58: 162-168.

Luo, P., Mu, D., Xue, H., et al., 2018a. "Flood inundation assessment for the Hanoi Central Area, Vietnam under historical and extreme rainfall conditions." *Scientific Reports* 8(1): 12623.

Luo, P., Zhou, M., Deng, H., et al., 2018b. "Impact of forest maintenance on water shortages: Hydrologic modeling and effects of climate change." *Science of the Total Environment* 615: 1355-1363.

Ma, D., Ding, Q., Li, Z., et al., 2012. "Prototype of an aquacultural information system based on internet of things E-Nose." World Automation Congress, Puerto Vallarta, Jalisco, July 24.

McGreavy, B., Webler, T., Calhoun, A.J., 2012. "Science communication and vernal pool conservation: A study of local decision maker attitudes in a knowledge-action system." *Journal of Environmental Management* 95(1): 1-8.

Miller, J.W., 2009. Farm ponds for water, fish and livelihoods. Food and Agriculture Organization of the United Nations (FAO).

Ministry of Agriculture of the PRC, 2009. Manual of producing and discharging coefficient for livestock and poultry from the first national census of pollution sources (restricted data).

Mitsuo, Y., Tsunoda, H., Kozawa, G., et al., 2014. "Response of the fish assemblage structure in a small farm pond to management dredging operations." *Agriculture Ecosystems & Environment* 188: 93-96.

Mushet, D.M., Calhoun, A.J.K., Alexander, L.C., et al., 2015. "Geographically isolated wetlands: Rethinking a misnomer." *Wetlands.* 35(3): 423-431.

Naigaga, S., Boyd, C.E., Gaillard, P., et al., 2017. "Assessing the reliability of water-test kits for use in pond aquaculture." *Journal of the World Aquaculture Society* 48(4): 555-562.

Ongley, E.D., Xiaolan, Z., Tao, Y., 2010. "Current status of agricultural and rural non-point source pollution assessment in China." *Environmental Pollution* 158(5): 1159-1168.

Osbaldiston, and R., Schott, J.P., 2012. "Environmental sustainability and behavioral science: Meta-analysis of proenvironmental behavior experiments." *Environ. Behavior* 44(2): 257-299.

Peng, S.Z., Xiong, Y.J., Luo, Y.F., et al., 2013. "The effect of paddy eco-ditch and wetland system on nitrogen and phosphorus pollutants reduction in drainage." *Journal of Hydraulic Engineering* 44(6): 657-663.

Perumal, T., Sulaiman, M. N., Leong, C. Y., 2016. "Internet of Things (IoT) enabled water monitoring system." 2015 IEEE 4th Global Conference on Consumer Electronics (GCCE), Osaka, Japan, February 4.

Poschlod, P., and Braun-Reichert, R., 2017. "Small natural features with large ecological roles in ancient agricultural landscapes of Central Europe-history, value, status, and conservation." *Biological Conservation* 211: 60-68.

Raebel, E.M., Merckx, T., Feber, R.E., et al., 2012. "Multi-scale effects of farmland management on dragonfly and damselfly assemblages of farmland ponds." *Agriculture Ecosystems & Environment* 161: 80-87.

Rains, M.C., Leibowitz, S.G., Cohen, M.J., et al., 2016. "Geographically isolated wetlands are part of the hydrological landscape." *Hydrological Processes* 30(1): 153-60.

Rashford, B.S., Bastian, C.T., Cole, J.G., 2011. "Agricultural land-use change in prairie Canada: Implications for wetland and waterfowl habitat conservation." *Canadian Journal of Agricultural Economics/Revue canadienne d'agroeconomie* 59(2): 185-205.

Schulze, J., Martin, R., Finger, A., et al., 2015. "Design, implementation and test of a serious online game for exploring complex relationships of sustainable land management and human well-being." *Environmental Modelling & Software* 65(3), 58-66.

Shogren, J.F., Parkhurst, G.M., Settle, C., 2003. "Integrating economics and ecology to protect nature on private lands: Models, methods, and mindsets." *Environmental Science & Policy* 6(3): 233-242.

Spiteri, A., and Nepalz, S.K., "2006. Incentive-based conservation programs in developing countries: A review of some key issues and suggestions for improvements." *Journal of Environmental Management* 37(1): 1-14.

Stocker, T., 2014. *Climate change 2013: The physical science basis: Working*

Group I contribution to the fifth assessment report of the Intergovernmental Panel on Climate Change. Cambridge: Cambridge University Press.

Tan X.M., 2005. *Hstory of irrigation and flood control in China.* Beijing: China Water Power Press.

Tang, Y.L., 2015. "Wetland conservation and construction strategies in the Leqing City." *East China Forest Management* 3: 38-40.

Thomsen, P.F., Kielgast, J., Iversen, L., et al., 2012. "Monitoring endangered freshwater biodiversity using environmental DNA." *Molecular Ecology* 21(11): 2565-2573.

Tian, L., Yuan, M.Q., Ma, Q.Y., et al., 2011. "An adaptive design for high stalk bulrush harvester in Southern China." *Journal of Central South University of Forestry & Technology* 31(11): 153-156.

Tiner, R.W., Lang, M.W., Klemas, V.V., 2015. *Remote Sensing of Wetlands: Applications and Advances.* Boca Raton: CRC Press.

UN-Water (United Nations World Water Assessment Programme), 2018. The United Nations World Water Development Report 2018: Nature-Based Solutions for Water. Paris, UNESCO.

Varela-Losada, M., Vega-Marcote, P., Pérez-Rodríguez, U., et al., 2016. "Going to action? A literature review on educational proposals in formal environmental education." *Environmental Education Research* 22(3): 390-421.

Verhoeven, J.T., Arheimer, B., Yin, C., et al., 2006. "Regional and global concerns over wetlands and water quality." *Trends in Ecology & Evolution* 21(2): 96-103.

Wang, G.H, Bai, F., Sang, W.G., 2017. "Spatial distribution of invasive alien animal and plant species and its influencing factors in China." *Plant Science Journal* 35(4): 513-524.

Wu, Q., 2018. "2.07-GIS and remote sensing applications in wetland mapping and monitoring." Comprehensive Geographic Information Systems 140-157.

Yan, X., Zhen, Y.L., Gregg, W.P., et al., 2001. "Invasive species in China-An overview." *Biodiversity and Conservation* 10(8): 1317-1341.

Yang, J., and Chen, W., 2017. "Behavioral Strategies of the Third-Party in Rural Community Building: A Case Study of Chenzhuang in Maoshan Mountain Scenic Area." *Modern Urban Research* 1: 18-22.

Yin, C., and Shan, B., 2001. "Multi-pond systems: A sustainable way to control diffuse phosphorus pollution." *AMBIO* 30(6): 369-375.

Yu, K.J., Jiang, Q.Z., Wang, Z.F., et al., 2015. "The Research progress and prospect of Beitang landscape." *Areal Research and Development* 34(3): 130-136.

Zhang, F., 2009. *History of irrigation engineering in ancient China.* Taiyuan: Shanxi Education Press.

Zhang, K., Shi, H., Peng, J., et al., 2018. "Microplastic pollution in China's inland water systems: A review of findings, methods, characteristics, effects, and management." *Science of The Total Environment* 630: 1641-1653.

Zhao, X., 2006. "Assessment of water environment and sustainable wetland development in the Zhalong Nature Reserve." *Water Resources Protection* 27(3): 22-25.

Zheng, H., Hong, Y., Long, D., et al., 2017. "Monitoring surface water quality using social media in the context of citizen science." *Hydrology & Earth System Sciences* 21(2): 949-961.

Zhou, B., Tan, X.M., Li, Y.P., et al., 2016a. "Quebei irrigation works and value analysis." *China Rural Water and Hydropower* 9: 57-61.

Chapter

③

Assessment of Nitrogen Pollution in a Pond-rich Agricultural Watershed

3.1 Introduction

Nitrogen (N) is crucial to aquatic and terrestrial organisms, and its use as a fertilizer in modern agriculture is one of the main reasons that crop production has kept pace with the human population growth (Robertson and Vitousek 2009). N is also a major pollutant facing humanity today, since excessive N discharge originated from intensive urban and agricultural sources can lead to eutrophication of lakes and reservoirs, groundwater nitrate enrichment, loss of biodiversity, and other serious environmental and human health issues (Filoso et al. 2004; Galloway et al. 2008; Chen et al. 2016). To reduce the adverse impacts of N pollution, many countries have incorporated water pollution mitigation strategies, such as the Total Maximum Daily Load program and various Best Management Practices, into watershed management. However, to develop suitable nutrient management programs requires knowledge of spatiotemporal variations of N concentrations in rivers and streams, as well as the load contributions of various pollution sources (Nestler et al. 2011; Liu et al. 2013; Shen et al. 2014; Hashemi et al. 2016).

River drainage networks are hierarchically organized systems where first- and second-order streams, usually referred to as headwater streams, constitute at least 70% of the total stream length (Lassaletta et al. 2010). In addition to their dominance in numbers and cumulative length, headwater streams also exert controls on stream runoff and downstream fluxes of nutrients. For example, Alexander et al. (2007) found that first-order headwaters contribute approximately 70% of the annual runoff and 65% of the N flux in second-order streams in the northeastern U.S. Dodds and Oakes (2008) estimated that water chemistry was closely correlated with riparian land cover adjacent to the first-order streams in eastern Kansas. These results are similar to the findings in European headwater agricultural watersheds (Lassaletta et al. 2010). However, owing to their small size and intermittent flows, headwater streams are often omitted or inaccurately represented on stream maps derived from topography, and are therefore ignored by water pollution mitigation strategies, in spite of their vulnerability to human activities and significant influences on downstream water quality (Armstrong et al. 2012; Rasmussen et al. 2013).

China is faced with the severe challenge of widespread eutrophication due to excessive discharge of nutrients, including N, especially for the southeastern region with high population density and rapid economic development. Many lakes and reservoirs in headwater regions have been designated as drinking water sources to provide sufficient amounts of high-quality water (Li et al. 2013). Although previous studies of N pollution of these waters have focused on agricultural drainage (Jeon et al. 2007; Liu et al. 2009; Shen et al. 2014; Hashemi et al. 2016), the pollution sources are usually more intricate (Wang et al. 2005; Ongley et al. 2010; Asian Development Bank 2011). For example, many small farms exist in the low mountains, where domestic sewage and waste are discharged to the environment without appropriate treatment. Besides, the improvement in people's living standards has driven growing appetites for meat and dairy products, but the burgeoning animal feeding operations usually lack sufficient waste disposal facilities in rural areas. Semi-continuous discharges of N pollutants from these mini-point sources can have a significant impact on both the watershed and downstream drinking water sources.

In addition to complex pollution sources, headwater agricultural watersheds are usually dominated by multi-pond systems in Southeastern China (Yin et al. 1993; Verhoeven et al. 2006; Chen et al. 2017). As a semi-natural wetland, a multi-pond system is composed of many tiny ponds, connected by ditches and streams. Such ponds have existed for centuries, serving as drinking water sources for farmers and livestock, and for washing, fishing, and irrigating. Recent studies show that multi-pond systems effectively retain NPS pollutants through sedimentation, adsorption, and uptake by aquatic plants (Liu et al. 2009). For instance, a five-year monitoring study in the Liuchahe Watershed near Chaohu Lake estimated that the connected ponds reduced phosphorus in agricultural drainage by > 90%, if the total pond area occupied 6%-10% of the entire watershed (Yin et al. 2006). Reduction capacities also show a gradient from foothill ponds to riverside ponds (Qiang et al. 2006). However, the dynamics of many small water bodies are challenging to quantify in terms of their relationship to complex pollution sources, and further, it is difficult to develop suitable nutrient management programs.

Many NPS pollution models with various capabilities and degrees of complexity have been developed in the past decades, including simple export coefficient models like PLOAD, regression models like SPARROW, and process-based models, including HSPF, SWAT, and INCA (Yang and Wang 2010; Butcher et al. 2014; Wellen et al. 2015). The simple models are commonly used to estimate pollution loads from different sources, but they are difficult to validate, and fail to account for pollutant migration and transformation (Ma et al. 2011; Liu et al. 2013; Lu et al. 2013). In contrast, process-based models have increasingly been applied to evaluate NPS pollution and associated uncertainties. In this study, HSPF was selected due to open-source characteristics, flexible input/output configurations, and the ability to simulate the fate and transport processes from various pollution sources to the final receiving waters. Nevertheless, HSPF has mainly been used to analyze nutrient discharge from agricultural drainage (Filoso et al. 2004; Jeon et al. 2007; Ribarova et al. 2008; Liu and Tong 2011; Li et al. 2015), despite its ability to incorporate different mini-point sources and surface waters (Bicknell et al. 2005; Yang and Wang 2010). Poor characterization of pollution sources and scattered small water bodies may lead to overestimation of the relative contributions from agriculture and disproportionate targeting of potential pollution sources during process-based simulations (Lassaletta et al. 2010).

The Chenzhuang (CZ) Watershed in the south of the Mt. Mao region was selected for this study as a testbed. It lies in the headwater region of two important basins with an environmental concern: the Taihu Lake Basin and Qinhuai River Basin. Although there are many small reservoirs designated as drinking water sources in peripheral areas, about 46% of them failed to reach the regulatory thresholds for N concentrations in the Chinese Surface Water Quality Standard (GB 3838-2002), which garners increased attentions from decision-makers and stakeholders (Li et al. 2013; Wang et al. 2016; Chen et al. 2017). In view of the above scientific challenges and the severe water quality situation in the reservoirs, the objectives of this work are to (1) employ HSPF to simulate N discharge and transport processes from all known anthropogenic mini-point and nonpoint pollution sources in the headwater agricultural watershed, (2) analyze the spatiotemporal patterns of N concentration in the multi-pond system, and quantify their load contributions from each pollution source, (3) compare the results with previous studies on N source attribution, and (4) propose suitable strategies to improve water quality for both the watershed and downstream regions. This study is a first attempt to incorporate multi-pond systems into the process-based modeling of NPS pollution. The proposed methods and analytical results can inform other hydro-environmental studies which focus on scattered and small water bodies in urban and natural systems. The methods and results are also useful to water pollution prevention and mitigation for entire river basins.

3.2 Materials and methods
3.2.1 Study area and data collection

Located upstream of the Lita Reservoir, the CZ Watershed covers an area of 4.8 km^2, and belongs to the subtropical monsoon climate zone with the mean annual temperature of 15.2 ℃ and the mean annual precipitation of 912 mm (Fig. 3.1). The watershed is surrounded by mountains and hills on three sides, rising from 62 to 295 m. There are 82 rural households and a registered population of 267 in the central area. The farmers take paddy-wheat crops and unfenced feeding animals as a source of food, and tea and beech trees as a source of income. With an area of 200 to 16,000 m^2 and an average depth of 1.5 to 3.4 m, the ponds are scattered around the village and connected by ditches for irrigation and agricultural drainage. Although industry is legally restricted in the Mt. Mao region, the local waters receive excessive nutrients from synthetic fertilizers, animal feces, and domestic sewage and waste. The Lita Reservoir is designed to achieve Level III Water in the GB 3838-2002 Standard, but the annual average TN concentration exceeds the threshold of Level V Water, which restricts the water uses from drinking and aquaculture to irrigation and scenery only (Chen et al. 2017).

Fig. 3.1 Map of the CZ Watershed showing the location, satellite image, watershed boundary, and monitoring sites.

Both streamflow and water quality were monitored from late 2013 to early 2017. The monitoring sites No. 1 to 3, No. 4 to 14, and No. 15 were in ditches, ponds, and stream/reservoir transition areas, respectively. At a monthly interval, water samples were collected, preserved, and analyzed for TN, NH$_3$-N, and NO$_x$-N concentrations, according to the national standard analytic methods for surface water and wastewater (China Environmental Protection Administration 2002). Meanwhile, water temperature and dissolved oxygen levels were measured using the YSI handheld water quality meter. Six typical sites in the ditches and pond outlets with continuous, year-round flows were chosen for streamflow monitoring, as national hydrometric stations are not located in such rural areas. Using Parshall flumes and LS1206B portable flow meters, streamflow data was collected once a week during the dry season (October to April), and every day during the wet season (May to September). In addition, daily weather data, including precipitation, humidity, solar radiation, air temperature, and wind direction/speed, was obtained by the WE800 meteorological station installed in the village. According to Wang et al. 2015, atmospheric N deposition was set as monthly constants, ranging from 0.2 to 0.9 kg·km^{-2} in dry days, and from 0.5 to 2.7 kg·km^{-2} in wet days.

3.2.2 HSPF description and data pre-processing

HSPF was developed by the US EPA as a deterministic, lumped-parameter, and comprehensive watershed model that continuously simulates land surface and subsurface hydrologic and water quality processes (Bicknell et al. 2005). Using the concept of hydrologic response unit (HRU), HSPF divides the watershed into three types of homogeneous segments: pervious land segment (PERLND), impervious land segment (IMPLND), and stream/reach/reservoir (RCHRES). It generates runoff and associated pollutants from land segments and routes the flow downslope, over intervening land segments and into a stream reach, and then from one reach to another downstream to the watershed outlet. Recent studies have shown that the performance of HSPF is comparable to other commonly used watershed models, such as SWAT and INCA (Yang and Wang 2010; Butcher et al. 2014; Wellen et al. 2015), but the loosely coupled submodules and configurable biogeochemical processes make it promising for application in small watersheds with various pollution sources and surface waters.

The employed modules and submodules in this study are outlined in Fig. 3.2. The outflow from land segments is calculated by PWATER and IWATER, which adopt empirical equations for interception, evaporation, infiltration, and groundwater loss processes. The fate and transport of N pollutants from land segments are simulated by PQUAL and IQUAL, which use accumulation and depletion rates to adapt to different land use patterns. Hydraulic behavior in the streams is processed by HYDR, which relies on the user-defined storage-volume relationships in FTABLE. The organic and inorganic N in receiving waters is calculated by NUTRX, following the subroutines of water temperature in HTRCH, dissolved oxygen and biochemical oxygen demand in OXRX, and plankton populations in PLANK. Besides, Special Actions are performed during the course of a model run to simulate human interactions with land segments and stream reaches. Further details on the above processes are described in

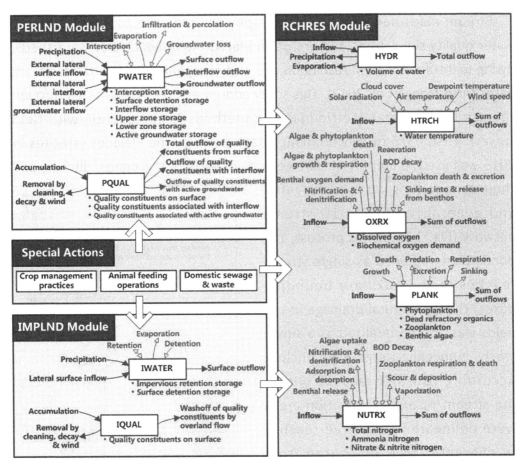

Fig. 3.2 Employed modules and submodules of HSPF, in which the blue, red, green and brown terms represent the major input, output, biogeochemical processes and variables in each submodule, respectively.

Donigian (2002), Bicknell et al. (2005), and other studies (Filoso et al. 2004; Jeon et al. 2007; He and Hogue 2012; Li et al. 2015; Ouyang et al. 2015).

HSPF relies on the Better Assessment Science Integrating Point and Nonpoint Sources (BASINS) to improve modeling efficiency (Duda et al. 2012), which contains a geographic information system program (MapWindow), a meteorological and time series data utility (WDMUtil), and an interactive Windows interface (WinHSPF). On the basis of these software tools, the data pre-processing of the study area proceeds as follows.

3.2.2.1 Water bodies generalization and segmentation

Stream maps derived from topography are requisite to hydrologic and water quality simulations, but are often inaccurate in headwater watersheds, owing to insufficient DEM resolution and anthropogenic changes to the stream paths (Armstrong et al. 2012). This study addresses these weaknesses through the use of photogrammetric mapping methods in conjunction with field surveys. First, a 0.5 m panchromatic image taken by the Pléiades satellites in 2015 was used to identify all surface waters. Second, the ponds, ditches, and their connections were investigated and generalized into a 1D stream network (Fig. 3.3). In this process, the ponds were regarded as single stream reaches in the auxiliary bounding boxes. The agricultural drainage in the fields was also generalized as a single reach, connecting to the main ditches. According to water surface width, the stream/reservoir transition areas were delineated into three reaches to characterize N discharge from upstream. Finally, FTABLEs containing depths, surface areas, volumes, and

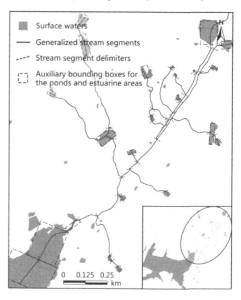

Fig. 3.3 Generalization and segmentation of the multi-pond system

flow information were prepared for all the generalized stream reaches.

3.2.2.2 Pervious and impervious land segmentation

Running HSPF requires the division of the watershed into several homogeneous land segments with varying physical and chemical attributes. The land segments are generated by overlaying subwatershed boundaries, a soil map, and a land use map. The physical attributes, including topsoil thickness (H), bulk density (ρb), soil erodibility factor (K), and percentage of pervious area (P), and the chemical attributes, such as pollutant concentrations in the topsoil, must be specified and assigned to each land segment.

Based on the TauDEM tools of BASINS and a 5-m resolution DEM from the Jiangsu Provincial Bureau of Surveying and Mapping, the subwatersheds of the study area were delineated by trial and error, when the derived channels and the above stream network were well matched (Fig. 3.4a). The three soil types, including Typic Hapli-Stagnic Anthrosols (THSA), Typic Claypani-Udic Argosols (TCUA), and Typic Hapli-Ustic Cambosols (THUC) (Fig. 3.4b), and their properties (Table 3.1) were mainly collected from Zhang et al. (1986). Other than the H and values, the K values of the three soil types were determined following Wischmeier and Smith (1978). The land use map was developed through semi-automatic feature extraction on the Pléiades image. In addition to the waters, the CZ Watershed consists of paddy fields (15.1%), dry farmland (9.8%), tea garden (3.6%), grove (12.4%), rural residential areas (2.9%), and woodland (51.4%) (Fig. 3.4c). The P value for the rural residential areas was set to 42% based on a field survey, and 100% for other land use types, according to Bicknell et al. (2005). Fourteen combinations of soil and land use types were obtained, after overlaying the soil and land use maps. Pollutant concentrations in the topsoil were calculated based on the H and ρb values, as well as laboratory analysis on the soil samples collected in each soil and land use combination (Table 3.2).

Fig. 3.4 Subwatersheds (a), soil types (b), land use types (c), and mini-point pollution sources (d) of the study area.

Table 3.1 Properties of the three soil types in the CZ Watershed

Soil types	Topsoil thickness (H, m)	Bulk density (ρb, g·cm⁻³)	Soil erodibility factor (K)
Typic Hapli-Stagnic Anthrosols (THSA)	0.29	1.24	0.04
Typic Claypani-Udic Argosols (TCUA)	0.16	1.26	0.03
Typic Hapli-Ustic Cambosols (THUC)	0.18	1.31	0.04

Table 3.2 Pollutant concentrations in the topsoil of the CZ Watershed

Soil and land use types	TN (kg·ha⁻¹)	NH₃-N (kg·ha⁻¹)	NOₓ-N (kg·ha⁻¹)	Soil and land use types	TN (kg·ha⁻¹)	NH₃-N (kg·ha⁻¹)	NOₓ-N (kg·ha⁻¹)
THSA, paddy field	2,960.32	29.54	6.49	TCUA, tea garden	1,909.14	13.29	1.51
THSA, dry farmland	2,327.41	23.37	9.57	TCUA, grove	1,591.27	14.87	1.62
THSA, grove	2,158.32	17.62	7.23	TCUA, woodland	1,127.27	13.34	3.19
THSA, woodland	2,766.89	24.32	5.26	TCUA, rural residential areas	962.38	8.87	1.39
THSA, rural residential areas	1,345.17	9.07	2.51	THUC, dry farmland	1,968.30	13.82	3.26
TCUA, paddy field	1,782.11	12.32	4.24	THUC, grove	1,326.28	9.68	2.06
TCUA, dry farmland	1,497.32	11.71	3.37	THUC, woodland	1,429.08	11.08	2.70

3.2.2.3 Crop management practices

Crop management practices need to be developed in as much detail as possible, especially with regard to crop rotations and fertilizer applications, as they demonstrate the effects of human interactions with the watershed, and are crucial to determine accurate N yields (Hashemi et al. 2016). Four main crop types, including paddy, wheat, tea, and beech tree, were discriminated and considered separately when running HSPF, although they were not fully delineated from the satellite image in Fig. 3.4c. According to face-to-face interviews with 21 local farmers, double paddy-cropping and paddy-wheat rotation are performed in paddy fields and dry farmland, whereas tea and beech trees are cultivated all year round. The schedules of annual cultivation and management practices for the crop types are shown in Table 3.3, in which the number of fertilizers and related nutrient concentrations were estimated based on regional statistics and each crop's agricultural manuals (Zhu and Liu 2003; Li et al. 2010; Yang and Fang 2015).

Table 3.3 Schedules of annual cultivation and agricultural practices

Crop type	Date	Operation	Crop type	Date	Operation
Paddy	Apr. 12/Aug. 3	Seeding	Tea	Oct. 27	Planting
	May 2/Aug. 21	Apply ammonium bicarbonate		Nov. 13	Apply compound fertilizer
	May 28/Sep. 17	Apply compound fertilizer		Mar. 2	Apply ammonium
	Jun. 24/Oct. 13	Apply urea		Mar. 21/Jul. 13	Tea-picking
	Jul. 20/Nov. 5	Grain harvesting		May 26/Aug. 29	Apply urea

Continued

Crop type	Date	Operation	Crop type	Date	Operation
	Aug. 16	Seeding		Mar. 10	Transplanting
	Aug. 27	Apply compound fertilizer		Apr. 1/Sep. 8	Intermediate cutting
Wheat	Nov. 9	Apply urea	Beech tree	Apr. 28	Apply compound fertilizer
	Feb. 13	Apply urea		Jul. 12	Apply compound fertilizer
	May 9	Grain harvesting		Oct. 14	Apply ammonium bicarbonate

3.2.2.4 Mini-point pollution sources

Since HSPF is more oriented towards biogeochemical processes, rather than point-source emissions, a simple export coefficient model (Johnes 1996) was employed to estimate N yields from animal feeding operations, rural domestic sewage and waste. Therefore, pollutant production coefficients for each individual resident and animal in conjunction with the scale and monthly variation of each pollution source were prepared.

Based on field surveys, twelve flocks of unfenced poultry (chicken and duck), three herds of unfenced livestock (goat), and six dumping areas of domestic waste were located on the map (Fig. 3.4d). The TN, NH_3-N, and NO_x-N production coefficients for each animal's feces and each resident's sewage and waste referred to our previous studies in the CZ Watershed (Chen et al. 2017), which was estimated according to a literature review focused on peripheral areas, and disaggregated into monthly constants. With a baseline number of 16 to 42 for each flock and 8 to 26 for each herd, the average monthly animal population decreased from 68% to 32% during autumn and winter, after breeding to 162% during spring and summer. Pollutants from waste dumps were discharged into the waters according to rainfall-runoff processes with the scales decided by the nearby resident population. Domestic sewage flows directly into the nearest ponds and ditches with the amount determined by the population in each household, which increases by 36 to 165% from the registered number during Chinese traditional festivals.

3.2.2.5 Meteorological, hydrological, and water quality database

Meteorological time series were managed using the WDMUtil tools of BASINS. HSPF requires eight meteorological time series to simulate the hydrological cycle in a watershed, which are precipitation, evaporation, air temperature, wind speed, solar radiation, potential evapotranspiration, dew point temperature, and cloud cover (Bicknell et al. 2005). In addition to the daily weather data collected from the meteorological station from late 2013 to early 2017, another four years of data (2010 to 2013) were obtained from the closest national climate station (the Liyang Station, No. 58345; 31°26'N, 119°29'E) for the warm-up period. The potential evapotranspiration was estimated by the Penman-Monteith equation (Allen et al. 1998). The dew point temperature was obtained from relative humidity and mean air temperature (China Meteorological Administration 2004). The cloud cover was calculated based on the actual sunlight duration and the possible sunshine duration for the latitude (Hamon et al. 1954).

The streamflow and water quality measurements were also integrated into the WDM database. For each monitoring site, the monthly streamflow was estimated based on weekly data, measured seven times during the dry season, and the sum of the daily measurements during the wet season. Daily streamflow was organized by the original measurements with empty values in the sequence. Monthly water quality measurements consisted of water temperature, dissolved oxygen levels, and TN, NH_3-N, and NO_x^--N oncentrations.

3.2.3 Sensitivity analysis, calibration, and validation

HSPF employs many parameters to describe watershed characteristics, and it is labor-intensive to make manual adjustments and almost impractical to derive the most suitable value for each parameter when HSPF is applied (Tang et al. 2006). Model-independent parameter estimation and uncertainty analysis software (PEST) was therefore employed to accelerate the traditional subjective process of sensitivity analysis and calibration (Doherty 2015). PEST applies an iterative process based on nonlinear estimation technique to adjust parameters one at a time. Local and relative sensitivities are simultaneously evaluated using Jacobean derivative-based composite measures, and then classified into several levels based on their Jenks natural breaks, so as to help identify critical parameters.

In this study, the parameters associated with the employed modules and recommended by the BASIN/HSPF Training Materials were first selected for sensitivity analysis (Donigian 2002; Bicknell et al. 2005; Duda et al. 2012). Their initial value ranges were obtained from previous HSPF applications and the computation increment was set as one-fiftieth of the initial range to limit the number of iterations in PEST. Then, using the derived parameters at the top sensitivity levels, the streamflow simulations were calibrated followed by the oxygen, plankton, and N simulations. Additionally, the parameters with high sensitivity values were set differently for ponds, ditches, and stream/reservoir transition areas, since the hydrological and N circles are determined by their various water retaining capacities, benthic oxygen demands, and rates of denitrification. Similarly, the parameters with apparent temporal variability, such as the empirical coefficients to quantify nutrient accumulation and washoff potency for land segments, were set differently for each month to further optimize the calibration.

The performance of HSPF was evaluated by qualitative and quantitative assessments. The qualitative procedures consisted of a visual comparison of the observations and simulations. The coefficient of determination (R^2) and the Nash-Sutcliffe efficiency coefficient (NSE) (Nash and Sutcliffe 1970) were used as the objective functions (see Eqs. (1) and (2)) to quantitatively evaluate model performance.

$$R^2 = 1 - \frac{\sum_{i=1}^{n}(S_i - f_i)^2}{\sum_{i=1}^{n}(S_i - \bar{S})^2} \tag{1}$$

$$NSE = 1 - \frac{\sum_{i=1}^{n}(O_i - S_i)^2}{\sum_{i=1}^{n}(O_i - \bar{Q})^2} \tag{2}$$

where n is the total number of observations, S_1 and O_1 are the simulated and observed values at the i–th time-step, $\bar{S}$ and $\bar{Q}$ are the mean of the simulated and observed values, and f_i is the value predicted by the linear regression equation between S_i and O_1.

Ranging from 0 to 1, R^2 indicates the proportion of the variance in measured data explained by the model. NSE is a normalized statistic that describes the degree of the "goodness-of-fit" between predictions and observations, and can vary between $-\infty$ and 1, where the value of 1 represents a perfect fit. According to several studies (Donigian 2002; Liu and Tong 2011), for monthly streamflow simulations, HSPF should achieve an R^2 value of 0.85, 0.75, and 0.65 to be considered "very good" "good" and "fair". For daily streamflow simulations, they are "very good" when the R^2 values are above 0.8, "good" when they are above 0.7, and "fair" when they are above 0.6. For monthly N concentration simulations, the three R^2 thresholds are 0.7, 0.6, and 0.5.

3.2.4 Source attribution of N pollution

In order to analyze the pollution source contributions, all known anthropogenic mini-point and nonpoint sources were first classified into six categories: food crop production, cash crop production, animal feeding operations, rural domestic sewage, rural domestic waste, and sediment release. Food crops referred to paddy and wheat, and cash crops included tea and beech trees. Aquaculture was not covered in this study as the waters were predominantly used for irrigation. Then, multiple runs of the calibrated and validated HSPF model with different pollution input scenarios were conducted to estimate the N load from each individual pollution source.

In the baseline scenario, HSPF was run without any of the six pollution sources, except the initial N constituents in the topsoil of land segments, denoted as $Load_{base}$. For the first five pollution sources, different model runs were then performed to estimate the combined N loads from initial soil constituents and each individual pollution source, whose difference from the baseline scenario was calculated as the load from that pollution source. For instance, when estimating the N load from food crop production, HSPF was run with crop management practices on paddy and wheat, so that the combined N load from initial soil constituent and food crop production, denoted as $Load_{base+food}$, was obtained. N discharge from food crop production could then be calculated as $Load_{base+food}$-$Load_{base}$. The same procedures were repeated to calculate the load contributions from cash crop production, animal feeding operations, rural domestic sewage, and rural domestic waste. N load from sediment release was calculated based on inner emissions from ponds, ditches, and stream/reservoir transition areas, which were obtained from the output variable of each stream reach when the other five pollution sources were added in a real scenario.

3.3 Results and discussion
3.3.1 Calibrated parameters and modeling performance

Sensitivity analysis highlighted the 26 most important parameters at the top four sensitivity levels for the streamflow and N simulations, with their categories, explanations, initial value ranges, and sensitivity values in Table 4. The subsequent HSPF calibration from January 2014 to December 2015 and validation from January to December 2016 were performed based on these parameters after the warm-up period from 2010 to 2013. Among the hydrologic parameters, UZSN, LZSN, and DEEPER were first calibrated using monthly streamflow at the six monitoring sites in the ditches and pond outlets. Then, INFILT, LZETP, and AGWRC were calibrated using the daily high and low flows, and INTFW and IRC were used to adjust the amount and timing of peak flows.

The difference between our calibrated values and the other HSPF applications in small agricultural watersheds (Donigian 2002; Liu and Tong 2011; Hayashi et al. 2015) mainly reflects parameters related to surface runoff and interflow. For example, for monthly streamflow calibration, our larger LZSN value led to lower surface runoff potential, especially in the paddy fields and dry farmland with ponds and ditches. For daily streamflow calibration, our smaller INFILT value and larger IRC value enabled more interflow storage, which would increase the contribution of interflow in total runoff, and cause more delay for peak flows to reach the stream/reservoir transition areas. The above differences were presumably due to the multi-pond system, which significantly affects the rainfall-runoff patterns by retaining water during continuous precipitation, and maintaining sufficient streamflow for irrigation during the dry season.

Table 3.4 Employed parameters and the calibrated results for the CZ Watershed

Category	Parameter	Initial range[a]	Sensitivity (S)	S level[b]	Calibrated value
	UZSN: upper zone nominal storage (mm)	1.20-25.50	0.140	II	4.20
	LZSN: lower zone nominal storage (mm)	50.00-320.00	0.282	III	260.00-310.00[c]
	DEEPFR: fraction of groundwater inflow to deep recharge	0.00-0.50	0.308	I	0.15
Hydrologic parameters	INFILT: soil infiltration rate (mmh^{-1})	0.00-1.75	0.411	IV	0.25-0.48[c]
	LZETP: lower zone evapotranspiration parameter	0.10-1.50	0.084	II	0.75
	AGWRC: groundwater recession rate (day^{-1})	0.80-0.99	0.136	I	0.64
	INTFW: interflow inflow parameter	1.00-3.00	0.165	III	1.00-1.32[c]
	IRC: interflow recession rate	0.00-0.90	0.113	II	0.80

Continued

Category Parameter		Initial range[a]	Sensitivity (S)	S level[b]	Calibrated value
Oxygen parameters	KBOD20: BOD decay rate at 20°C (h^{-1})	0.05-0.15	0.027	I	0.10
	KODSET: rate of BOD settling (mh^{-1})	0.00-0.30	0.011	I	0.18
	BENOD: benthal oxygen demand at 20°C (mgm^{-2}·day^{-1})	0.20-5.20	0.202	III	0.30-3.20[c]
Plankton parameters	MALGR: maximal algal growth rate for phytoplankton (h^{-1})	0.00-0.80	0.124	II	0.24
	ALR20: unit respiration rate at 20°C (h^{-1})	0.00-0.60	0.046	I	0.36
	PHYSET: rate of phytoplankton settling (mh^{-1})	0.00-0.50	0.086	II	0.15-0.26[c]
Nitrogen parameters	ACQOP: accumulation rate of quality constituent on the surface (kg·ha^{-1}·day^{-1})	0.00-0.50	0.119	II	0.15
	SQOLIM: maximum storage of quality constituent on the surface (kg·ha^{-1}·month^{-1})	0.00-10.00	0.477	IV	3.80-6.20[c]
	WSQOP: rate of surface runoff that will remove 90% of stored quality constituent (m·h^{-1})	0.10-4.20	0.316	IV	1.45-2.80[c]
	MON-POTFW: monthly washoff potency factor of quality constituent	1.00-20.00	0.172	III	4.40-8.26[d]
	MON-INFLW-CONC: monthly concentration of quality constituent in interflow (mg·m^{-3})	0.10-3.00	0.476	IV	1.15-2.12[d]
	MON-GRND-CONC: monthly concentration of quality constituent in active groundwater (mg·m^{-3})	0.10-3.00	0.085	II	0.90
	KTAM20: unit oxidation rate of total ammonia at 20°C (h^{-1})	0.01-0.60	0.131	II	0.34

Category Parameter		Initial range[a]	Sensitivity (S)	S level[b]	Calibrated value
Nitrogen parameters	KNO220: unit oxidation rate of nitrite at 20 C (h⁻¹)	0.01-0.30	0.274	III	0.06-0.16[c]
	TCNIT: temperature correction coefficient for the nitrogen oxidation rates	1.03-1.07	0.030	I	1.04
	KNO320: unit denitrification rate of nitrate at 20 C (h⁻¹)	0.01-0.60	0.196	III	0.28-0.45[c]
	TCDEN: temperature correction coefficient for the denitrification rate	1.01-1.15	0.073	II	1.02
	BRNIT2:benthal release rate of inorganic nitrogen under anaerobic conditions (mg·m⁻²·h⁻¹)	−8.00-40.00	0.206	III	−1.28-12.16[c]

[a] The initial value range of the hydrologic parameters referred to Liu and Tong (2011), He and Hogue (2012), Hayashi et al. (2015), and Li et al. (2015), while the oxygen, plankton and N parameters referred to Filoso et al. (2004), Ribarova et al. (2008), Yang and Wang (2010), Li et al. (2015), and Ouyang et al. (2015).

[b] The range of sensitivity value for each level: I, 0.000-0.060; II, 0.060-0.150; III, 0.150-0.300; IV, 0.300-0.500.

[c] The range of the calibrated values for different land segments or stream reaches.

[d] The range of the calibrated values for different months.

Following the hydrologic calibrations, oxygen, plankton, and N parameters were evaluated using monthly water quality data. Among the empirical coefficients associated with N accumulation and removal, ACQOP, SQOLIM, WSQOP, and MON-POTFW were first estimated considering the TN, NH_3-N, and NO_x^--N constituents in the topsoil, and MON-INFLW-CONC and MON-GRND-CONC were then determined via water quality measurements. Similar to other N simulations using HSPF (Filoso et al. 2004; Yang and Wang 2010; Li et al. 2015), SQOLIM, WSQOP, and MON-INFLW-CONC were the most sensitive parameters, but our calibrated values were slightly larger, likely due to the accumulated N pollutants from mini-point sources in the study area. In addition, according to the calibrated results of the biogeochemical parameters in different stream reaches, such as KNO220 and KNO320, N retention and removal were more significant in ponds and stream/reservoir transition areas than in ditches, likely due to their various capacities for water storage and nutrient uptake.

Table 3.5 lists the overall performance of HSPF for simulating streamflow and N concentrations. According to the results at six streamflow monitoring sites, HSPF satisfactorily simulated monthly and daily streamflow with R^2 values above 0.70, and NSE values mostly above 0.65. However, there were some differences for the monthly N simulations. The R^2 and NSE statistics for the TN and NH_3-N concentrations ranged between 0.65 and 0.78, much higher than the range from 0.51 to 0.67 for the NO_x-N concentrations. According to the criteria suggested by Donigian (2002) and Liu and Tong (2011), HSPF achieved "very good" and "good" results for streamflow, TN, and NH_3-N simulations, whereas the performance of NO_x-N simulations can only be considered "fair".

Table 3.5 HSPF modeling performance during the calibration and validation period.

Simulations	R^2		NSE	
	Calibration	Validation	Calibration	Validation
Monthly streamflow	0.85	0.81	0.80	0.74
Daily streamflow	0.74	0.71	0.62	0.68
TN concentrations	0.78	0.74	0.70	0.71
NH_3-N concentrations	0.73	0.69	0.75	0.65
NO_x-N concentrations	0.67	0.61	0.57	0.51

To further verify the performance of the model, Fig. 3.5 shows a visual comparison between the observed and simulated N concentrations at three typical stream reaches for the period 2014–2016. As Fig. 3.1 shows, the sites No. 5, No. 2, and No. 15 represent an upstream pond near the tea garden, the main ditch next to the village, and the downstream transition areas, respectively. The results indicated that observations and simulations generally agree, but temporal variations were different for the three sites. TN concentrations showed high midyear variability and low levels at the end of the year, whereas the concentrations at sites No. 5 and No. 2 were generally larger than those at site No. 15. For the NH$_3$-N concentrations, it was similar to the TN variations at site No. 2, but more stable and higher at site No. 5, and lower at site No. 15. The NO$_x^-$-N concentrations were lower at sites No. 5 and No. 15, but higher at site No. 2. These variations were possibly due to variable pollutant discharge from the land segments, and the N mitigation and transformation in the multi-pond system that was identified by the calibrated HSPF.

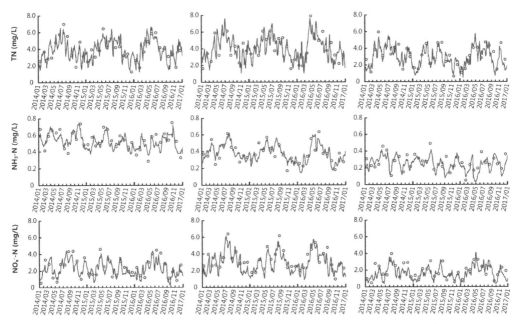

Fig. 3.5 Comparison of the observed and simulated N concentrations at three typical stream reaches

3.3.2 Spatiotemporal analysis and source attributions of N pollution

Using the calibrated and validated HSPF model, average seasonal N simulations of each stream reach are presented in Fig. 3.6. For TN concentrations, 52 to 100% of the ponds and ditches exceeded Level V Water in the GB 3838-2002 Standard throughout the year, whereas the estuarine areas exceeded the threshold by a factor of two in spring and summer. The NH$_3$-N concentrations reached Level III Water overall, but exhibited a 43% increase for the ponds near the village in summer. For NO_x^--N concentrations, waters far from the village mostly exceeded Level V Water in spring, summer, and autumn, which were 1.26, 1.47, and 1.19 times the concentrations near the village. The overall seriousness of the N indicators was therefore ranked as: TN > NO_x^--N > NH_3-N in the study area. The TN pollution was severe for the entire watershed throughout the year, and the NO_x^--N pollution was significant for the ponds and ditches far from the village.

Fig. 3.6 Average seasonal variations and spatial distribution of the simulated N concentrations. [a] The first five ranges refer to the levels in the GB 3838-2002 Standard, whereas the last two ranges represent exceeding the threshold of Level V Water by a factor of one and two respectively

According to the above spatiotemporal analysis, the scattered waters were divided into five classes to further quantify the pollution contributions from various mini-point and nonpoint sources. The five classes include the PNV (Ponds Near the Village), PFV (Ponds Far from the Village), DNV (Ditches Near the Village), DFV (Ditches Far from the Village), and DTA (Downstream Transition Areas) (Fig. 3.7a). The average seasonal N loads discharged from land segments into the waters were estimated by multiplying the N concentrations by the monthly streamflow, and subtracting the load discharge from upstream reaches (Fig. 3.7b). It was estimated that all the waters received the largest amount of N loads in summer, whereas the PFV alone received 32%, 38%, and 52% of the total TN, NH_3-N, and NO_x^--N loads throughout the year, clearly surpassing the other four classes. The N loads in the PNV and DTA were close to each other, but they were 5.4 and 3.3 times the loads in the DNV and DFV on average. The DNV and DFV received the lowest N loads in all seasons, but more pollutant discharges were identified far from the village.

Fig. 3.7 Water body classification (a) and their average seasonal N loads (b). [a] The five water classes are divided by the three colored dashed lines in (a).

Average monthly TN, NH_3-N, and NO_x-N loads in the five water classes were used to analyze the contributions from various pollution sources. For TN loads, food crop production and animal feeding operations were the largest contributors in the PNV, accounting for 34% and 24%, respectively, whereas cash crop production, rural domestic sewage and waste, as well as sediment release contributed less than 10% all year-round. In the PFV, cash crop production, animal feeding operations, and food crops were major contributors, accounting for 32%, 33%, and 19%, respectively, whereas the other three sources dropped to 5%. While cash crop production and animal feeding operations dropped to 9%, rural domestic sewage and waste, as well as food crop production were still major contributors, accounting for 30%, 23%, and 21%, respectively, in the DNV. Cash crop production rose as the main contributor at 32%, and food crop production, animal feeding operations, rural domestic sewage and waste contributed about 16%, respectively, in the DFV. Pollution source attributions in the DTA were similar to that in the PFV, but sediment release was much higher at 19% (Fig. 3.8a).

Fig. 3.8 Average monthly N loads from various pollution sources that discharged into the five water classes of the CZ Watershed

Source attributions of the NH$_3$-N loads were slightly different from the TN loads. Rural domestic sewage, animal feeding operations, and food crop production were significant contributors for the PNV, accounting for 28%, 27%, and 24%, respectively, whereas the other three sources contributed about 21% altogether throughout the year. As the largest contributor, cash crop production accounted for 41% of the NH$_3$-N load in the PFV, followed by food crop production and animal feeding operations, which contributed about 16%, respectively. Pollution source attributions in the DNV were similar to that in the PNV, where rural domestic sewage, animal feeding operations, and food crop production accounted for 38%, 24%, and 20%, respectively, and the other sources contributed about 18% altogether. In the DFV, cash crop production still ranked as the largest contributor, accounting for 39%, whereas food crop production, animal feeding operations, and rural domestic sewage contributed about 17%, respectively. The NH$_3$-N attributions are evenly distributed for the DTA, in which the first group, including cash crop production, animal feeding operations, and rural domestic sewage, accounted for 65% of pollution altogether, and the second group, including the other three sources, contributed about 35% (Fig. 3.8b).

The composition of pollution sources for the NO$_x$-N loads was very different from the TN and NH$_3$-N loads. Food and cash crop production were dominant contributors for all five water classes, accounting for 52% and 24% in the PNV, 31% and 49% in the PFV, 48% and 24% in the DNV, 40% and 45% in the DFV, and 36% and 38% in the DTA throughout the year. Animal feeding operations contributed a small proportion of 13% and 15% for the PNV and DNV, whereas the other sources accounted for less than 10%, respectively, in all waters (Fig. 3.8c).

3.3.3 Comparisons with previous studies on N source attribution

Previous process-based modeling studies on N pollution mostly focused on agricultural drainage, with some consideration to loads from municipal and industrial point sources, but few considered pollutants from rural life and animal feeding operations, as well as their adverse effect on small water bodies in headwater watersheds. For example, Li et al. (2015) evaluated the performance of HSPF in predicting runoff and nutrient export at the outlet of a typical small watershed in the hilly eastern monsoon region of China. Their results indicated that TN concentrations tend to be higher in subwatersheds with intensive agricultural activities. Lam et al. (2010) employed SWAT to evaluate the long-term influence of point source (urban and industrial wastewater) and diffuse source (fertilizer application and atmospheric deposition) pollution on the ecological status of the North German lowlands. Their results pointed to agriculture as the dominant contributor of NO_x^--N loads for the river mainstem. Using an extended SWAT model, Shen et al. (2014) identified paddy fields and dry farmland as the major source of TN and NH_3-N loads with decreasing trends from upstream to downstream tributaries and highest variations during the cultivation season in the Upper Reach of the Yangtze River.

In contrast, quite a few N load studies use empirical coefficient methods and encompass more types of pollution sources and surface waters. For instance, Ma et al. (2011) employed the simple export coefficient model to investigate water quality in the streams and creeks of Hubei Province in Central China. Their findings demonstrated that agricultural land use, rural domestic waste, and animal feeding operations accounted for 31%, 22%, and 12% of the TN loads, respectively. Lu et al. (2013) used an inversed Bayesian statistical approach to estimate TN export coefficients from commonly available stream monitoring data. They estimated that fertilizer applications and rural residential sewage contributed 47% and 29% of the TN loads in headwater tributaries of the Changle River watershed in Southeastern China. Liu et al. (2013) developed a systematic empirical framework to estimate nutrient release from five major sectors of the Taihu Lake Basin in Eastern China. They concluded that rural life, fertilizer applications, and animal feeding operations accounted for 46%, 19%, and 7% of the TN loads, respectively. Despite the differences among their watershed characteristics and modeling methods, crop production was identified as a major contributor of TN loads in rivers, streams, and lakes, while rural life and animal feeding operations reflected smaller contributions.

Several studies also employ an intermediate method between the export coefficient models and process-based models to quantify N source attributions. For instance, Chen et al. (2010) combined a nutrient balance module with an emission inventory module to analyze the nutrient emissions from the China agro-ecosystem. They provided an entire picture of agricultural N pollution at the national level, in which livestock feeding, fertilizer applications, rural life, and crop waste disposal accounted for 50%, 34%, 13%, and 3%, respectively. Ding et al. (2010) improved the export coefficient model by integrating the precipitation and terrain impact factor to delineate the spatiotemporal heterogeneities of nutrient discharge. They estimated that dry farmland, animal feeding operations, rural life, and paddy fields contributed 27%, 13%, 8%, and 5% respectively of dissolved N loads in the Upper Reach of the Yangtze River. Wang et al. (2011) established a dualistic structure model by integrating pollutant transport coefficients and natural/social correction factors into empirical export coefficients. They concluded that agricultural fields alone had contributed approximately 77% of the TN loads in the Yangtze River Basin, which was quite different from the other intermediate modeling studies.

Using the process-based HSPF model to incorporate a local multi-pond system and all known anthropogenic pollution sources, we identified crop production as the major source of N pollution, in which food and cash crop production had the largest impact on waters near and far from the village. TN loads from animal feeding operations ranked third in all ponds, whereas ditches were more vulnerable to rural residential sewage, and stream/reservoir transition areas were seriously affected by sediment release. With regard to the NH_3-N loads, rural residential sewage and animal feeding operations were identified as the third and fourth pollution contributor to the ponds and ditches. As for the NO_x-N loads, pollution attribution from the other four sources was far less than crop production for all waters. Using a 1D stream network, the multi-pond system was generalized and incorporated into the process-based modeling of NPS pollution for the first time. This setting offered more detailed analytic results compared to previous studies based purely on stream maps. For instance, owing to the pollutant retention and removal in the ponds far from the village, the simulated NH_3-N and NO_x^--N concentrations tended to be higher in the outlet ditch than in the inlet ditch of those ponds (Fig. 3.6). Meanwhile, mini-point pollution sources, including animal feeding operations and rural domestic sewage and waste, were identified based on field surveys. Our pollution source attribution generally agreed with results from studies that employed export coefficient and intermediate modeling methods, resulting in an average proportion of 47% from mini-point sources for TN and NH_3-N loads (Fig. 3.8). These results indicate that the mini-point sources are non-negligible contributors of N pollution in headwater agricultural watersheds.

Since the sources of N in nitrate and suspended particulate matter vary in their isotopic signatures, isotope techniques with various sampling strategies and interpretation models have been used frequently to quantify N pollution attribution in surface and ground water (Nestler et al. 2011; Denk et al. 2017). In China, Li et al. (2014) used a multi-isotopic method (^{15}N, ^{18}O, and ^{13}C) together with water chemistry to elucidate the source and fate of nitrate in the Miyun Reservoir and its upstream rivers. Their results indicated that soil N, sewage and animal feces were the major sources of nitrate pollution in the rivers, whereas a higher contribution from atmospheric N deposition appeared in the reservoir. Wang et al. (2016) combined the dual isotope approach with a Bayesian mixing model to evaluate diffusive nitrate sources in the headwater agricultural watersheds in the Jianghuai Hilly Region. They estimated that soil N and fertilizer application contributed about 63% of nitrate pollution during the wet season, compared to 22% from sewage and animal feces. However, during the dry season, soil N and fertilizer application accounted for 49%, compared to 33% from sewage and animal feces. Owing to the limited monitoring data on atmospheric N deposition, we employed measurements collected in peripheral areas (Wang et al. 2015), and identified similar phenomena of dominant contributions from soil N and fertilizer application, and increasing contributions from sewage and animal feces from the wet season to the dry season (Fig. 3.8).

3.3.4 Implications to headwater agricultural watersheds

Over the past decades, eutrophication and algal blooms have occurred in many lakes and rivers worldwide. In this context, the headwater watersheds with close ties to downstream drinking water sources were regarded as the last areas free from water quality deterioration (Lassaletta et al. 2010; Li et al. 2013). Nevertheless, most of the waters in the CZ Watershed failed to reach the national standard of Level V Water for TN and NO_x^--N concentrations, although there are no intensive pollution point sources, such as industrial wastewater. NH_3-N concentrations met the standard of Category III Water overall, but the ponds near the village were still contaminated in summer (Fig. 3.6). These unexpected results underscore the impact of agricultural headwaters on downstream water quality, and suggest the need for immediate pollution control strategies with an emphasis on TN and NO_x^--N loads.

China is widely acknowledged to be the largest consumer of synthetic N fertilizers. Lack of training in fertilizer applications and poor knowledge of soil properties has led small farmers to misuse fertilizers with consequent N losses into the environment (Ongley et al. 2010; Yang and Fang 2015; Chen et al. 2016). In view of their role as the largest contributor of N pollution, developing eco-fertilizer programs to reduce excessive nutrient discharge should be a top priority in headwater watersheds. Additionally, poultry and livestock feeding are employed to meet the requirement for meat and dairy in rural areas, but scattered and uncontrolled production makes it difficult to reduce adverse environmental impacts (Wang 2005; Asian Development Bank 2011). Animal feeding operations are therefore encouraged to organize into concentrated and fenced areas with regular collection of feces and wastewater. In addition, although the installation of sewage treatment facilities and garbage collection systems is feasible, technologies for converting sewage and waste into biogas and organic fertilizers are preferred (Lim et al. 2016). If designed and used properly, these sustainable methods can reduce N loads, and help close the nutrient loop. With a close tie to rural life, all three categories of mini-point sources were tracked across the five water classes, so their associated N discharge was characterized as intricate and widespread as fertilizer application. In order to effectively target these pollutants, we suggest training courses and mutual help for small farmers to build environmental awareness and move them toward water pollution mitigation, shifting from "end-of-pipe" treatment to source prevention and control.

Multi-pond systems serve a crucial role in irrigation, animal feeding operations, and rural residential water use. Although their water retention capacities can reduce flood peaks and nutrient loads, a large amount of N pollution accumulates in the sediments of these scattered ponds and ditches, and becomes a potential source for both the watershed and downstream drinking water sources. In this study, the ponds were generalized as a single stream reach in HSPF (Fig. 3.3). However, the hydro-environmental processes of those semi-natural wetlands are more complex. For example, intermittent overland flow into the ponds is mainly determined by surrounding micro-topography, which is difficult to delineate from either satellite imagery or field surveys (Yin et al. 2006; Chen et al. 2017). Moreover, rates of biogeochemical reactions, such as reaeration and denitrification, are heterogeneous in different locations of each pond (Verhoeven et al. 2006), which brings high uncertainty to the quantification of nutrient adsorption and uptake. Therefore, in order to better understand the source-sink dynamics of a multi-pond system, we recommend experiments on sediment release, detailed modeling of N retention and discharge during rainfall-runoff processes, and establishing user-friendly tools to help identify the most polluted and threatened waters.

3.4 Conclusions

As critical components of large river basins, headwater agricultural watersheds, characterized by scattered and small water bodies, are vulnerable to multiple mini-point and nonpoint pollution sources. Knowledge of the spatiotemporal patterns of N concentration, and individual load contribution from each pollution source is requisite to develop suitable nutrient management programs. We employed the process-based HSPF model to estimate N pollution and source attribution for a local multi-pond system in the CZ Watershed in the Mt. Mao Region of Southeastern China. Major findings of this study include the following:

(1) HSPF yielded "very good" and "good" results for streamflow, TN, and NH_3-N simulations, but "fair" for the NO_x^--N simulations. The calibrated parameters and modeling performance demonstrate that HSPF is capable of describing the particular characteristics of a multi-pond system and mini-point pollution sources.

(2) N pollution exhibited a distinctive spatial seasonal pattern with an overall seriousness rank: TN > NO_x^--N > NH_3-N. TN pollution was severe for the entire watershed. NO_x^--N pollution was significant for the waters far from the village. NH_3-N concentrations were acceptable except for the ponds near the village in summer.

(3) Food and cash crop production was identified as the largest source of N loads. However, mini-point pollution sources, including the combined effects of animal feeding operations, rural residential sewage and waste, accounted for as much as 47% of the TN and NH_3-N loads in ponds and ditches.

(4) Developing eco-fertilizer programs, concentrated animal feeding operations, and sustainable sewage treatment facilities should be prioritized for water quality improvement. Environmental awareness building for small farmers is also indispensable.

Water quality deterioration due to excessive N discharge is an issue of concern worldwide. To formulate the most suitable mitigation strategies, many studies have been conducted on the estimation and source attribution of N pollution at various spatiotemporal scales. Centered on a typical headwater watershed in the low mountains, the incorporation of the multi-pond system and mini-point sources in this study offers more details in the process-based modeling of NPS pollution. The results are also useful to sustainable agricultural development and water quality improvement for entire river basins. Owing to the limited geomatic and monitoring data, ponds were generalized as single stream reaches in HSPF. In the future, observations with higher spatiotemporal resolutions, such as LiDAR data from UAVs (10 cm point-spacing), real-time water quality monitoring (30 min. time-spacing), and in-situ experiment on sediment, can be used to better identify the hydrological patterns of scattered and small water bodies, their source-sink dynamics during rainfall-runoff processes, and detailed pollution source attribution with associated uncertainties.

References

Alexander, R. B., Boyer, E. W., Smith, R. A., et al., 2007. "The role of headwater streams in downstream water quality." *Journal of the American Water Resources Association* 43(1): 41-59.

Allen, R. G., Pereira L. S., Raes D, et al., 1998. "Crop evapotranspiration-Guidelines for computing crop water requirements-FAO Irrigation and drainage paper 56." *FAO.* 300(9): D05109.

Armstrong, A., Stedman, R. C., Bishop, J. A., et al., 2012. "What's a stream without water? Disproportionality in headwater regions impacting water quality." *Journal of Environmental Management* 50: 849-860.

Asian Development Bank, 2011. Nonpoint source pollution control in catchment areas. 44040012.

Bicknell, B., Imhoff, J., Kittle, J., et al., 2005. "Hydrological Simulation Program-FORTRAN: HSPF version 12.2 user's manual." *Water science and technology : A journal of the International Association on Water Pollution Research* 56.

Butcher, J.B., Johnson, T.E., Nover, D., et al., 2014. "Incorporating the effects of increased atmospheric CO2 in watershed model projections of climate change impacts." *Journal of Hydrology* 513: 322-334.

Chen, M., Chen, J., Sun, F., 2010. "Estimating nutrient releases from agriculture in China: An extended substance flow analysis framework and a modeling tool." *Science of the Total Environment* 408(21): 5123-5136.

Chen, M., Sun, F., Shindo, J., 2016. "China's agricultural nitrogen flows in 2011: Environmental assessment and management scenarios." *Resources, Conservation and Recycling* 111(4): 10-27.

Chen, W., Duan, W., He, B., et al., 2017. "Water quality modeling for typical rural watershed based on the WASP model in Mountain Mao Region, upper Taihu Basin." *Journal of Lake Science* 29(4): 836-847.

China Environmental Protection Administration, 2002. *Water and wastewater monitoring and analysis method.* Beijing: China Environmental Science Press.

China Meteorological Administration, 2004. *Specifications for surface meteorological observation.* Beijing: China Meterological Press.

Denk, T.R., Mohn, J., Decock, C., et al., 2017. "The nitrogen cycle: A review of isotope effects and isotope modeling approaches." *Soil Biology and Biochemistry* 105: 121-137.

Ding, X., Shen, Z., Hong, Q., et al., 2010. "Development and test of the export coefficient model in the upper reach of the Yangtze River." *Journal of Hydrology* 383(3): 233-244.

Dodds, W.K., and Oakes, R.M., 2008. "Headwater influences on downstream water quality." *Journal of Environmental Management* 41(3): 367-377.

Donigian, A.S., 2002. "Watershed model calibration and validation: The HSPF experience." *Proceedings of the Water Environment Federation* 2002(8): 44-73.

Duda, P.B., Hummel, P.R., Donigian, A.S., et al., 2012. "BASINS/HSPF: Model use, calibration, and validation." *Transactions of the Asabe* 55(4): 1523-1547.

Filoso, S., Vallino, J., Hopkinson, C., et al., 2004. "Modeling nitrogen transport in the Ipswich River basin, Massachusetts, using a Hydrological Simulation

Program in FORTRAN (HSPF)". *Journal of the American Water Resources Association* 40(5): 1365-1384.

Galloway, J. N., Townsend, A. R., Erisman, J, W., et al., 2008. "Transformation of the nitrogen cycle: recent trends, questions, and potential solutions." *Science* 320(5878): 889-892.

Hashemi, F., Olesen, J. E., Dalgaard, T., et al., 2016. "Review of scenario analyses to reduce agricultural nitrogen and phosphorus loading to the aquatic environment." *Science of the Total Environment* 573: 608-626.

Hayashi, S., Murakami, S., Xu, K.Q., et al., 2015. "Simulation of the reduction of runoff and sediment load resulting from the Gain for Green Program in the Jialingjiang catchment, upper region of the Yangtze River, China." *Journal of Environmental Management* 149: 126-137.

He, M., and Hogue, T. S., 2012. "Integrating hydrologic modeling and land use projections for evaluation of hydrologic response and regional water supply impacts in semi-arid environments." *Environmental Earth Sciences* 65(6): 1671-1685.

Jeon, J.H., Yoon, C. G., Donigian, A. S., et al., 2007. "Development of the HSPF-Paddy model to estimate watershed pollutant loads in paddy farming regions." *Agricultural Water Management* 90(1): 75-86.

Johnes, P. J., 1996. "Evaluation and management of the impact of land use change on the nitrogen and phosphorus load delivered to surface waters: The export coefficient modelling approach." *Journal of Hydrology* 183(3-4): 323-349.

Lam, Q. D., Schmalz, B., Fohrer, N., 2010. "Modelling point and diffuse source pollution of nitrate in a rural lowland catchment using the SWAT model." *Agricultural Water Management* 97(2): 317-325.

Lassaletta, L., García-Gómez, H., Gimeno, B. S., et al., 2010. "Headwater streams: Neglected ecosystems in the EU Water Framework Directive. Implications for nitrogen pollution control." *Environmental Science & Policy* 13(5): 423-433.

Li, H., Zhang, W., Zhang, F., et al., 2010. "Analysis of the changes in chemical fertilizer use and efficiency of the main grain crops in China." *Plant Nutrition and Fertilizer Science* 16(5): 1136-1143.

Li, H., Zhu, G., Chen, W., et al., 2013. "Current situation of good water quality reservoirs in hilly region of south-east China: Protection practices of Tianmuhu Reservoir." *Journal of Lake Science* 25(6): 775-784.

Li, X., Liu, C., Liu, X., et al., 2014. "Sources and processes affecting nitrate in a dam-controlled subtropical river, Southwest China." *Aquatic Geochemistry* 20(5): 483-500.

Li, Z., Liu, H., Luo, C., et al., 2015. "Simulation of runoff and nutrient export from a typical small watershed in China using the Hydrological Simulation Program-Fortran." *Environmental Science and Pollution Research* 22(10): 7954-7966.

Lim, S. L., Lee, L. H., Wu, T. Y., 2016. "Sustainability of using composting and vermicomposting technologies for organic solid waste biotransformation: Recent overview, greenhouse gases emissions and economic analysis." *Journal of Cleaner Production* 111: 262-278.

Liu, B., Liu, H., Zhang, B., et al., 2013. "Modeling nutrient release in the Tai Lake Basin of China: Source identification and policy implications." *Environ Manage* 51(3): 724-737.

Liu, Y., Fu, Q., Yin, C., 2009. "Phosphorus sorption and sedimentation in a multi-pond system within a headstream agricultural watershed." *Water Quality Research Journal of Canada* 44(3): 243-252.

Liu, Z., Tong, S. T. Y., 2011. "Using HSPF to model the hydrologic and water quality impacts of riparian land-use change in a small watershed." *Journal of Environmental Informatics* 17(1): 1-14.

Lu, J., Gong, D., Shen, Y., et al., 2013. "An inversed Bayesian modeling approach for estimating nitrogen export coefficients and uncertainty assessment in an agricultural watershed in eastern China." *Agricultural Water Management* 116: 79-88.

Ma, X., Li, Y., Zhang, M., et al., 2011. "Assessment and analysis of non-point source nitrogen and phosphorus loads in the Three Gorges Reservoir Area of Hubei Province, China." *Science of the Total Environment* 412-413: 154-161.

Nash, J. E., and Sutcliffe, J. V., 1970. "River flow forecasting through conceptual models part I-A discussion of principles." *Journal of Hydrology* 10(3): 282-290.

Nestler, A., Berglund, M., Accoe, F., et al., 2011. "Isotopes for improved management of nitrate pollution in aqueous resources: Review of surface water field studies." *Environmental Science and Pollution Research* 18(4): 519-533.

Ongley, E.D., Zhang, X., Yu, T., 2010. "Current status of agricultural and rural non-point source pollution assessment in China." *Environmental Pollution* 158(5): 1159-1168.

Ouyang, Y., Leininger, T. D., Moran, M., 2015. "Estimating effects of reforestation on nitrogen and phosphorus load reductions in the Lower Yazoo River Watershed, Mississippi." *Ecological Engineering* 75: 449-456.

Qiang, F., Yin, C., Shan, B. 2006. "Phosphorus sorption capacities in a headstream landscape-The pond chain structure." *Journal of Environmental Sciences* 18(5): 1004-1011.

Ribarova, I., Ninov, P., Cooper, D., 2008. "Modeling nutrient pollution during a first flood event using HSPF software: Iskar River case study, Bulgaria." *Ecological Modelling* 211(1-2): 241-246.

Robertson, G. P., and Vitousek, P. M., 2009. "Nitrogen in agriculture: Balancing the cost of an essential resource." *Annual review of environment and resources* 34: 97-125.

Shen, Z., Qiu, J., Hong, Q., et al., 2014. "Simulation of spatial and temporal distributions of non-point source pollution load in the Three Gorges Reservoir Region." *Science of the Total Environment* 493: 138-146.

Tang, T., Reed, P., Wagener, T., et al., 2007. "Comparing sensitivity analysis methods to advance lumped watershed model identification and evaluation." *Hydrology and Earth System Sciences* 11(2): 793-817.

Verhoeven, J. T., Arheimer, B., Yin, C., et al., 2006. "Regional and global concerns over wetlands and water quality." *Trends in Ecology & Evolution* 21(2): 96-103.

Wang, M., Lu, B., Wang, J., et al., 2016. "Using dual isotopes and a Bayesian isotope mixing model to evaluate nitrate sources of surface water in a drinking water source watershed, East China." *Water* 8(8): 355.

Wang, X., 2005. "Diffuse pollution from livestock production in China." *Chinese Journal of Geochemistry* 24(2): 189-193.

Wang, X., Hao, F., Cheng, H., et al., 2011. "Estimating non-point source pollutant loads for the large-scale basin of the Yangtze River in China." *Environmental Earth Sciences* 63(5): 1079-1092.

Wang, Y., Liu, N., Wang, J., 2015. "Study on atmospheric deposition of nitrogen and phosphorus in Taihu Lake." *Environmental Science and Management* 40(5): 103-105.

Wellen, C., Kamran-Disfani, A. R., Arhonditsis, G. B., 2015. "Evaluation of the current state of distributed watershed nutrient water quality modeling." *Environmental Science & Technology* 49(6): 3278-3290.

Wischmeier, W., and Smith, D., 1978. Predicting rainfall erosion losses. Agricultural Handbook 537. Agricultural Research Service, United States Department of Agriculture.

Yang, X., and Fang, S., 2015. "Practices, perceptions, and implications of fertilizer use in East-Central China." *Ambio* 44(7): 647-652.

Yang, Y., and Wang, L., 2010. "A review of modeling tools for implementation of the EU water framework directive in handling diffuse water pollution." *Water Resources Management* 24(9): 1819-1843.

Yin, C., Shan, B., Mao, Z., 2006. "Sustainable water management by using wetlands in catchments with intensive land use." In: Verhoeven, J.T.A., Beltman, B., Bobbink, R., Whigham, D.F. (eds) *Wetlands and Natural Resource Management. Ecological Studies,* vol 190. Berlin: Springer.

Yin, C., Zhao, M., Jin, W., et al., 1993. "A multi-pond system as a protective zone for the management of lakes in China." *Hydrobiologia* 251(1): 321-329.

Zhang, Y., Zhou, M., Wu, X., 1986. Soil reconds of Jurong Country in Jiangsu Province. Soil Survey Office of Jurong County, Zhenjiang Department of Agriculture.

Zhu, T., and Liu, X., 2003. *Handbook of concise techniques on agricultural fertilization.* Beijing: Golden Shield Press.

Chapter

4

Quantifying the Hydrologic Regulation
Effects of Agricultural Multi-pond Systems

4.1 Introduction

Wetlands provide a portfolio of ecosystem services, from floodwater storage to pollutant removal, habitat provision, and microclimate regulation. These services are governed to a large extent by the hydrologic processes that occur within wetlands, and interactions with the catchment in which the wetlands are located (Cohen et al., 2016; Lane et al., 2018; Chen et al., 2019). Understanding the hydrologic processes and assessing their regulation effects over scales ranging from single wetland to regional is a historical challenge of earth system sciences (Rains et al., 2016; Ameli and Creed, 2017). These knowledges also serve as a prerequisite for developing best management practices (BMPs), low impact development (LID), and natural-based solutions (NBSs), when addressing a range of environmental and socioeconomic challenges toward sustainable development (UN-Water, 2018; UNEP, 2019).

Multi-pond systems (MPSs), traditional and semi-artificial water conservancy projects, are a common type of wetland in southern China (Yin and Shan, 2001; Verhoeven et al., 2006; Liu et al., 2009). An MPS typically consists of several ponds, scattered aside farmland and cottages (Fig. 4.1a). These ponds are natural surface depressions that permanently or temporarily hold water, and then gradually become dammed and partly connected by ditches to stabilize water storage. Such ponds are regular features of the landscape, owing to the low cost of construction and maintenance, high efficiency in irrigation and drainage, and value for multiple uses, including drinking, cleaning and fishing. First constructed in the 200s Before Common Era, MPSs are now widespread with an estimated total amount exceeding 6 million (Yu et al., 2015; distributions shown in Fig. 4.1b). Although we discuss the Chinese context, similar small, scattered wetlands with perennial or intermittent surface water connections are also reported in other humid agricultural regions, such as the North American Coastal Plain (Evenson et al., 2018b; Golden et al., 2019), the Mediterranean Basin (Poschlod and Braun-Reichert, 2017), and the Southern Highlands of Australia (Williams and Fryirs, 2016).

Fig. 4.1 A typical agricultural landscape with MPSs (a) and their distributions in southern China (b) (adapted from Chen et al., 2019)

In catchments with MPSs, rainfall initially fills shallow ponds and then runs off along the ditches to downgradient ponds and other waters. This process helps to reduce flood peaks, contributes to groundwater recharge, and provides baseflow for irrigation (Golden et al., 2014; Rains et al., 2016). For example, the 193 ponds were found to reduce annual irrigation water shortages from 306±26 to 89±48 mm in the Liuchahe catchment, Anhui Province (Yin et al., 2006). During a heavy rainfall event (141 mm d^{-1}), these ponds reduced the flood peak from 2.5 to 0.3 $m^3 s^{-1}$ (Liu et al., 2009). Although hydrologic regulation of these small, scattered wetlands has been widely and quantitatively recognized at catchment scales (Wang et al., 2008; Longbucco, 2010; Feng et al., 2013), three scientific questions remain regarding their intra-catchment processes: (1) how to specify the pond fill-spill relationship among the pervasive and interlaced ditch network, (2) to elucidate the structure of MPSs in which ponds are perennially or intermittently connected, and (3) to assess their hydrologic regulation over scales ranging from single pond to single MPS and entire catchment.

Recent advances in geospatial data and analytics have enabled process-based watershed models, such as Soil and Water Assessment Tool (SWAT), Hydrological Simulation Program-Fortran (HSPF), and Annualized Agricultural Non-Point Source Pollution Model (AnnAGNPS), to incorporate small, scattered wetlands (Golden et al., 2017; Chen et al., 2018a; Yasarer et al., 2018). Relative to lumped wetland parameter representation (Wang et al., 2008; Feng et al., 2013), the spatially-explicit representation with physical and functional attributes (e.g., submerged area and storage capacity) provides feasible means for detailed assessments on these waters (Evenson et al., 2015; Yasarer et al., 2018; Evenson et al., 2018b; Yeo et al., 2019). However, model limitations still exist for consideration of MPSs. Notable processes represented include bidirectional exchanges between ponded water and shallow groundwater (Rains et al., 2016), and cascading fill-spill relationships between intermittently connected ponds (Yu et al., 2015). The latter has been observed as an important hydrologic component, but is not well resolved in previous studies, owing to the difficulties in depicting the hydrologic path of ditches, even from high-resolution Digital Elevation Models (DEMs) and remote sensing imagery (Wu, 2018; Chen et al., 2019). Furthermore, the availability of measured data has limited further exploration on these intra-catchment hydrologic details, requiring both streamflow and pond dynamics for parameter estimation (Yen et al., 2016).

Over the past decades, small, scattered wetlands, such as MPSs, have faced a severe decline in numbers and malfunctions in terms of ecosystem services owing to expanding urbanization, agricultural intensification, and centralized irrigation (Hill et al., 2018; Chen et al., 2019). While little environmental legislation has focused on pond conservation, several policies and regulations have gradually acknowledged their importance, such as the headwater environment remediation in China's River Chief Mechanism that clarifies the responsibilities of different government sectors (Dai, 2019), and pond planning and dredging in Beautiful Rural Development that pursues both ecological and economic well-being (Long, 2014). Pond prioritization and management for environmentalists and rural planners also requires detailed assessments on the structures and processes of these waters to achieve the potential benefits of MPSs.

Given the above scientific challenges and management deficiencies, this study presents a novel modelling approach to explore the multi-scale hydrologic regulation of MPSs in humid agricultural catchments. The headwaters of two basins of environmental concern (the Taihu Lake Basin and Qinhuai River Basin, Fig. 4.1b) were selected as a test bed. A new version of SWAT (v. 2012, rev. 670) was developed to represent MPS (hereafter referred to as SWAT-MPS), so that intermittent spillage connections between ponds and their bidirectional exchanges with shallow groundwater can be better represented. SWAT-MPS was set up with both streamflow and pond water-level variations and calibrated using rule-based artificial intelligence. The simulations were conducted under two contrasting land-use scenarios to assess changes in the hydrologic components of ponds, MPSs, and the entire catchment, followed by discussions on their multi-scale hydrologic regulation, implications for management, and modelling recommendations. As a first modelling exploration on the intra-catchment hydrologic details of MPSs, the proposed methods and analytical results can better quantify the water transport and distribution in catchments with small, scattered wetlands. They can also help inform the management and conservation efforts in the context of global wetland degradation.

4.2 Materials and methods
4.2.1 Study area and data

The study catchment (31°42'N, 118°18'E) is characterized by traditional agricultural practices, densely distributed ponds, and a humid subtropical monsoon climate (Fig. 4.2a). A 5-m resolution DEM from the Jiangsu Provincial

Bureau of Surveying and Mapping shows that the 4.8 km² catchment is surrounded by low mountains and hills, rising from 62 to 295 m (Fig. 4.2b). According to Chen et al., 2018a, the soil types include Eutric Planosols, Haplic Luvisols, and Dystric Regosols (Fig. 4.2c), which are composed of silty loam and loam textures at the top soil layer, and sandy loam and clay loam textures below, with saturated hydraulic conductivity values of 9.2-210.5 and 2.1-38.8 mm hr⁻¹, respectively (calculated by Saxton and Rawls, 2005). In addition to agricultural land (paddy fields, dry farmland, and tea gardens totaling 29%), the catchment consists of woods (deciduous and evergreen forests at 63%), waters (ponds and streams at 5%), and rural residential areas (3%), as extracted from a 0.5-m panchromatic satellite image taken in 2017 (Fig. 4.2d).

Fig. 4.2 Study catchment with numbered ponds (a), DEM (b), land use (c), and soil classes (d). Note: the pervasive and interlaced ditch network exists between ponds and the main stream, but was only partially surveyed, nor shown in (a).

Two-year (June 2017 to May 2019) complete data including daily meteorological conditions and variations of streamflow, pond water-level and shallow groundwater table was prepared, while the former 2 categories were collected for more than these years. National Ground Climatological Data (Station No. 58345; N31°26', E119°29', 7.7 m in elevation) indicated average daily high and low temperatures at 21.1 and 13.3°C, respectively, while average annual precipitation of 1136±352 mm with ~ 60% concentrated from June to September for the past half-decade. Another preliminary analysis showed small variations (~ -8 to 3%, Fig. A4.1) in relative percentage changes on the average daily values of precipitation, maximum/minimum temperature, and sunshine duration of the 2 years to past half-decade. With fixed ground conditions including DEM, soil type, and land use, variations on these key meteorological variables of a watershed model indicated that the credibility of subsequent hydrologic patterns of test MPSs is beyond the 2-year period with complete data.

Daily streamflow at the catchment outlet was estimated using a pre-established HSPF model for the past 4 years (Chen et al., 2018a), which was calibrated with measurements of a portable propeller-type current meter and later a fixed Parshall flume in the midstream and validated with daily average Nash-Sutcliffe Efficiency (NSE) and $R^2 > 0.8$. Among the 30 ponds in total (area > 100 m^2), 9 were selected and gaged with the HOBO U20L for water-level observation, according to the Jenks natural breaks of pond elevation, storage capacity and submerged area. Manual irrigation pumping of these ponds was also investigated. Pond bathymetry was surveyed and integrated with the surrounding DEM topography, to derive the relationship between water-level, storage, and submerged area for each pond (see Chen et al., 2018b for detailed procedures). Furthermore, underlain by a confining unit ~30 m below, the peripheral regions have a shallow groundwater table ~6 m below the surface (Geological Bureau of Jiangsu Province, 1980). Water-level variations in 6 drinking water wells, which locate in and adjacent to the catchment, and are deep to the same aquifer, were therefore monitored for the estimation of shallow groundwater table, after establishing the Thiessen polygons of the wells and catchment boundary using ArcToolbox.

4.2.2 The SWAT-MPS model

4.2.2.1 Default wetland modelling in SWAT

SWAT is an open-source, semi-distributed, and semi-physically based watershed model that has been widely applied to examine the impacts of agricultural land management on water resources and water quality constituents (Arnold et al., 2012). Hydrologic simulation in SWAT is first performed at the hydrologic response unit (HRU, a combination of soil, land use and slope that determines rainfall-runoff characteristics) level, and then aggregated at the subwatershed and routed to the watershed outlet following the stream network. In this process, wetlands can be modelled at the subwatershed scale, in which all wetlands are depicted via aggregated or "lumped" parameters (Wang et al., 2008), or at the HRU scale, which enables a customized inflow proportion from a single HRU or subset of all HRUs for each wetland (Feng et al., 2013; Golden et al., 2014). The two modes share a common water balance equation, which is expressed as:

$$\Delta V = P - ET + Q_{in} - Q_{out} - Q_{seep} \tag{1}$$

where ΔV is the change in the wetland water volume during a given time-step (m^3), P is the water volume entering as precipitation (m^3), Q_{in} and Q_{out} are the water volume flowing in and out (m^3), and ET and Q_{seep} are the water volume lost via evapotranspiration and seepage (m^3). Further details on these parameters are described in Arnold et al. (2012), Kang et al., (2017), and Sridhar et al., (2019).

The SWAT default wetland model has substantial limitations, when assessing the intra-catchment hydrologic processes of MPSs. First, pond inflows are determined by empirical or constant parameters in both of the two modes, rather than their actual distribution and drainage characteristics, while pond outflows (i.e., spillage) are directly routed to streams, ignoring possible fill-spill relationships with downgradient ponds (Ameli and Creed, 2017; Jones et al., 2019). Second, pond subsurface outflows (i.e., seepage) are identified at a constant rate dictated by soil saturated hydraulic conductivity values and soil water content. This simplification precludes representation of changes in head gradients between these waters and surrounding uplands, which drive the direction and magnitude of local groundwater exchange (Foglia et al., 2007; Golden et al., 2014). The above limitations of original SWAT stem from the structural representation of HRU for ponds and their drainage areas, as well as process representation of the water balance equation that ignores the cascading fill-spill processes and bidirectional exchange with shallow groundwater.

4.2.2.2 Improved representation of MPSs

Significant efforts have been made to enhance process representation of wetlands in SWAT and to set up more spatially explicit wetland parameters. Relevant improvements include: (1) delineating upland-to-wetland drainage relationships and constraining wetland surface and subsurface inflows according to actual spatial positioning (Evenson et al., 2015; Yasarer et al., 2018; Yeo et al., 2019); and (2) supporting bidirectional exchange between wetland surface and subsurface water as a function of time-varying head gradients with the water content in soil profiles (Cohen et al., 2016; Evenson et al., 2018b). On this basis, SWAT-MPS was developed to better present MPSs by: (1) enabling pond-to-pond and pond-to-stream fill-spill relationships, especially for those between different subcatchments, as the intermittent water connections are determined by pond water storage and surrounding microtopography, but usually unidentifiable even from high-resolution geospatial datasets; and (2) estimating pond surface and subsurface water exchange according to head gradients between ponds and drinking water wells, as the latter are common, easily accessible, and can reflect the areal shallow groundwater table when deep to the same aquifer in low mountains and hills.

Following Evenson et al. (2015 and 2018b), the SWAT HRUs were first redefined as the combination of soil, land use, slope, and the fourth attribute of pond drainage area, so that the upland drainage relationship of each pond is spatially-explicit (Fig. 4.3a). In this process, the drainage area was depicted via the multiple flow direction approach (Tarboton, 2005), and new model input files were added to enhance the structural attributes of pond HRUs. Implemented by the matrix data structure in the Python scientific library, each pond can then receive spillage from upgradient ponds, and overflow to its subcatchment reach and downgradient ponds in and out of the same subcatchment, to restore an intermittent fill-spill relationship that may not be presented by subcatchment delineation (Fig. 4.3a). Furthermore, the relationship between water storage and submerged area was discretized at 0.1 m water-level intervals, while the total outflow magnitude was estimated via the flow calculation formula of broad-crested weirs (see Herschy, 2008 for detailed procedures) and surveyed practice of irrigation pumping for each pond.

The hydrologic processes of pond HRUs are extended and reimplemented to better represent the fill-spill relationship within an MPS and bidirectional

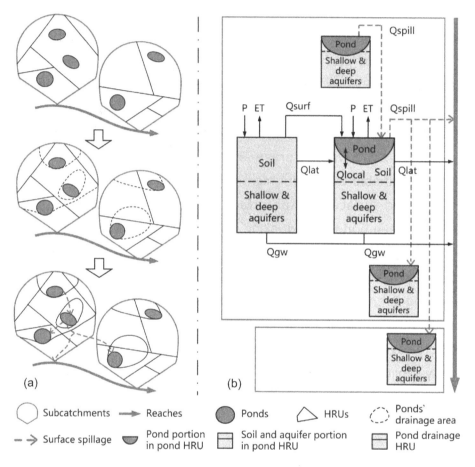

Fig. 4.3 Improved HRU structural (a) and process (b) representation in SWAT-MPS. The left figure shows (1) the redefinition of HRU boundaries based on default HRUs and the pond drainage areas following Evenson et al. (2015 and 2018b), and (2) the newly added cascading fill-spill relationships between ponds in and out of the same subcatchment. The right figure shows (1) the spillage received and provided by the pond portion of a pond HRU, i.e., Q_{spill} in Eq. (3) and Eq. (4), and (2) the bidirectional interacts with the soil and aquifer portion of a pond HRU, i.e., Q_{local} in Eq. (5). Other underlined hydrologic components of the pond HRU include precipitation (P), evapotranspiration (ET), surface runoff (Q_{surf}), lateral subsurface flow ($Q_{lateral}$), and groundwater flow (Q_{gw}).

exchange between pond surface water and shallow groundwater. Hence, pond HRUs are resolved into two parts: the pond portion (Fig. 4.3b, marked in blue) and soil and aquifer portion (Fig. 3b, marked in green). The water balance of the pond portion is described as:

$$\Delta V_{Pond} = P - ET + Q_{surf} - Q_{spill\,out} + Q_{spill\,in} \pm Q_{local} \qquad (2)$$

where ΔV_{Pond} is the change in the pond water volume during a given time-step (m^3). P and ET are the water volume entering as precipitation and lost via evapotranspiration (m^3) from the time-varying submerged area. Q_{surf} is the surface runoff (m^3) entering from pond drainage HRUs. Q_{spill_out} and Q_{spill_in} are the surface water (m^3) leaving and entering via the fill-spill processes, which are depicted through the following equations:

$$Q_{spill\_out} = \sum_i^m \left(p_{pond,i} \times Q_{spill\_out}\right) + \left(p_{stream} \times Q_{spill\_out}\right) \qquad (3)$$

$$Q_{spill\_in} = \sum_i^n \left(p_{pond,i} \times Q_{spill\_out}\right) \qquad (4)$$

where m and n are the total number of ponds that receive and provide spillage, respectively. $P_{pond,i}$ is the proportion of total spillage that routes to the-th pond, while P_{stream} is the proportion that routes to the subcatchment reach. Note that the summation of $\sum_i^m P_{pond,i}$ and P_{stream} equals 1. Q_{local} is the net subsurface flow (m^3) entering (+) and leaving (-) via local exchange with underlying soil and shallow aquifers, which is estimated by head gradients between ponds and drinking water wells according to Darcy's law (Sophocleous, 2002). Hence, Q_{local} is simulated as:

$$Q_{local} = K_{sat} \times S_{bottom} \times \sum_i^m \left(\frac{H_{pond} - H_{well,\,i}}{L_i} \times w_i\right) \qquad (5)$$

where K_{sat} is the saturated hydraulic conductivity (m day^{-1}) of the pond's substrate. S_{bottom} is the time-varying submerged area (m^2) on the integrated bathymetric and terrain surface. H_{pond} and $H_{well,\,i}$ are water-levels (m) in the pond and the i-th well, respectively. L_i is the distance (m) from the well to the pond's centroid, while w_1 is the inverse distance weight to that well according to the distance to all measured wells. The subsurface flow was depicted as entering from the underlying soil layer and shallow aquifers if the surrounding water table height exceeded the pond water-level ($H_{well} > H_{pond}$), and in turn, leaving the pond if the water-level exceeded the surrounding water table height ($H_{pond} > H_{well}$).

Beneath the pond portion, the soil layer and aquifer portion of pond HRU is simulated as:

$$\Delta V_{Beneath} = Q_{lat\_in} - Q_{lat\_out} \pm Q_{local} - Q_{gw} \qquad (6)$$

where $\Delta V_{Beneath}$ is the change in the water volume (m^3) of soil layer and aquifers in the pond HRU during a given time-step. Q_{lat_in} and Q_{lat_out} are the lateral subsurface flow (m^3) entering the soil profile from drainage HRUs and leaving the pond HRU to either a downgradient HRU or the subcatchment reach, respectively. Q_{gw} is the groundwater outflow (m^3) leaving shallow and deep aquifers and entering the reach.

4.2.3 Model calibration and validation

Daily streamflow in original SWAT was first prepared over 4 seasonal years (June 2015 to May 2019), after a 3-year warm-up period (June 2012 to May 2015) to get basic parameter ranges and catchment conditions. SWAT-MPS was then calibrated and validated against daily streamflow and pond water-level variations over the latest 2 years (June 2017 to May 2019). Here, seasons were defined as spring (March to May), summer (June to August), autumn (September to November), and winter (December to February), according to the subtropical climate. The calibration and validation practices include parameter sensitivity analysis, rule-based knowledge integration, Genetic Algorithm (GA)-based parameter optimization, cross-validation, and goodness-of-fit assessment (Fig. 4.4). The initial sensitivity analysis was executed using both one-at-a-time and global sensitivity analysis in the SWAT-CUP software package (Abbaspour, 2013). With full physically meaningful value ranges suggested by Arnold et al., 2012 and Abbaspour, 2013, the sensitive parameters of original SWAT in conjunction with the above new ones for pond hydrologic processes, i.e., P_{pond}, P_{stream}, and K_{sat}, are employed for the following procedures.

4.2.3.1 Rule-based fill-spill relationship

Calibrated by streamflow measurements, hydrologic models usually

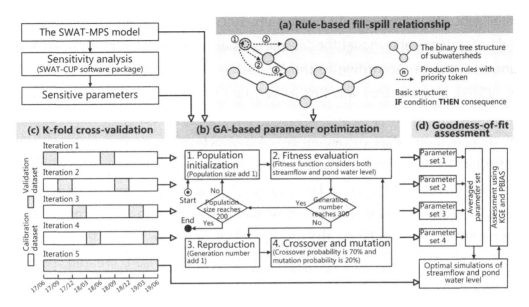

Fig. 4.4 Flowchart of SWAT-MPS calibration and validation

lack accurate intra-catchment simulations, despite satisfactory results at the catchment outlet (Efstratiadis and Koutsoyiannis, 2010). Two major difficulties emerged with respect to SWAT-MPS calibration and validation: (1) Overall rationality. Parameter estimation on a single pond cannot guarantee appropriate value sets, as the fill-spill processes may be associated with several upgradient and downgradient ponds. Meanwhile, the catchment rainfall-runoff and aggregated regulation of all MPSs jointly influence the outlet streamflow. (2) Uncertainties in the direction and magnitude of pond overflow. The cascading fill-spill processes of an MPS are affected by the ponds' low-lying terrain, the pervasive and interlaced ditch networks, microtopography in farmland, and various rainfall intensity (Yu et al., 2015; Chen et al., 2019). These factors are difficult to monitor and model, and are usually changed by agricultural practices, resulting in uncertainties in the direction and magnitude of pond overflow.

To cope with the above difficulties and narrow the value range of pond representation parameters, production rules, which originated from artificial intelligence (Engelmore and Feigenbaum, 1993), were introduced to integrate field experience and model the fill-spill possibility between pond-to-pond and pond-to-stream. According to the Strahler's Stream Order (Scheidegger et al., 1965), a binary tree was first established to present the hierarchical and topological relationship of all subcatchments (Fig. 4.4a). Using a basic structure of "**IF** condition **THEN** consequence", 4 production rules were designed and set priority tokens for possible overflow between ponds in single or multiple subcatchments (Table 4.1). Ordered by rule ID, the fill-spill processes are more likely to happen between ponds in subcatchments with a closer topological relationship. After replacing the downgradient pond (i.e., Pond$_j$ in Table 4.1) with the main stream in these rules, another 4 rules were created for the spillage between pond-to-stream, and in aggregate, implemented for the following parameter optimization.

Table 4.1 Production rules for the pond fill-spill relationship

ID	Content
Rule 1	**IF** Pond$_i$ belongs to Subcatchment$_m$ **AND** Pond$_j$ belongs to Subcatchment$_m$ **AND** Height of Pond$_i$ > Height of Pond$_j$ **THEN** Pond$_i$ can spill to Pond$_j$
Rule 2	**IF** Pond$_i$ belongs to Subcatchment$_m$ **AND** Pond$_j$ belongs to Subcatchment$_n$ **AND** Subcatchment$_m$ is **a brother node** of Subcatchment$_n$ **AND** Height of Pond$_i$ > Height of Pond$_j$ **THEN** Pond$_i$ can spill to Pond$_j$
Rule 3	**IF** Pond$_i$ belongs to Subcatchment$_m$ **AND** Pond$_j$ belongs to Subcatchment$_n$ **AND** Subcatchment$_m$ is **a father node** of Subcatchment$_n$ **AND** Height of Pond$_i$ > Height of Pond$_j$ **THEN** Pond$_i$ can spill to Pond$_j$
Rule 4	**IF** Pond$_i$ belongs to Subcatchment$m$ **AND** Pond$_j$ belongs to Subcatchment$_n$ **AND** Subcatchment$_m$ is **a grandfather node** of Subcatchment$_n$ **AND** Height of Pond$_i$ > Height of Pond$_j$ **THEN** Pond$_i$ can spill to Pond$_j$

4.2.3.2 GA-based parameter optimization

Genetic Algorithms (GA), originally proposed by Goldberg (1989), are increasingly used in parameter estimation of hydrologic modelling due to the advantages in global optimization (Shafii and Smedt, 2009; Chlumecký et al., 2017). Following sensitivity analysis and rule-based knowledge integration, the original SWAT and new pond representation parameters were subsequently optimized using a Python-based evolutionary computation framework against daily streamflow at the catchment outlet and water-level variations in the selected ponds. The GA optimization was performed in the following 4 steps, and then repeated 100 times to evaluate the stability of GA and acquire 20 optimal parameter value sets (Fig. 4.4b). After comparing with the parameter value ranges reported by Arnold et al., 2012, Abbaspour, 2013 and Lin et al., 2013, the most reasonable set was manually selected, considering both model performance and actual fill-spill relationships that have been partly surveyed in the field.

(1) **Population initialization.** Two hundred initial parameter sets (population) were generated for the first GA run. Parameters were encoded in the form of a substring and linked together to form one individual, while their values were randomly generated within the pre-defined ranges.

(2) **Fitness evaluation.** During the training period, the fitness (F) of each individual was evaluated by a fitness function based on the root-mean-square error (RMSE) of streamflow and pond water-level, which is described as:

$$F = \frac{1}{\sum_{i=1}^{n} \sqrt{\sum_{j=1}^{m} (P_{ij}^{obs} - P_{ij}^{sim})^2/m} + k\sqrt{\sum_{j=1}^{m} (S_{j}^{obs} - S_{j}^{sim})^2/m}} \tag{7}$$

where P_{ij}^{obs} and P_{ij}^{sim} are observed and simulated water-level for the i-th pond on the j-th day, respectively. S_{j}^{obs} and S_{j}^{sim} are observed and simulated streamflow on the j-th day. As a weight coefficient, k took the same value as n to balance the two types of measurement. Ranging from 0 to $+\infty$, a higher F value implied a lower RMSE value and higher fitness.

(3) **Reproduction** aimed to create the next generations by selecting individuals with a higher fitness value (F) and endowing them with a higher chance to reproduce. Hence, the overall model fit of the individuals is gradually improved over 300 generations.

(4) **Crossover and mutation** were expected to increase population diversity

and generate dissimilar individuals (parameter values) with the high model fit. For each generation, 65% of the individuals were randomly selected for the crossover process, which generated new individuals by slightly changing parameter values to derive the different magnitude of pond overflow. Meanwhile, mutation introduced random change to the parameter values of 25% of individuals, in which the above 8 production rules were implemented by probabilistic events to generate different spillage directions.

Further details on the GA implementation can be found in Shafii and Smedt (2009) and Fortin et al. (2012).

4.2.3.3 Cross-validation and goodness-of-fit assessment

A k-fold cross-validation was employed for model validation, owing to the relatively short period of measured data (Fig. 4.4c). The streamflow and water-level variations of the 2 years were first divided into 4 subsets according to the seasons. In each of the 4 cross-validation processes, 3 subsets were held for the above parameter optimization, while the remaining one was used for model validation. Hence, each subset of the measured data was used once for validation and 4 models were generated accordingly. Finally, the parameter values of the 4 models were averaged and performed on the complete data for the goodness-of-fit assessment. Further theoretical basis and calculation processes for the cross-validation are described in Foglia et al. (2007) and Ly et al. (2013).

The goodness-of-fit assessment of SWAT-MPS covered both streamflow and pond water-level simulations. Two statistical criteria, including the Kling-Gupta Efficiency (KGE) and Percent Bias (PBIAS), were employed to evaluate the agreements between simulations and measurements (Fig. 4.4d). As a derivative of the Nash-Sutcliffe Efficiency (NSE) statistic (Nash and Sutcliffe, 1970), the KGE reduces the inherent tendency of runoff underestimation in NSE, while allowing for a multi-objective perspective by focusing on the correlation error, variability error, and bias error (Guptaet al., 2009). A KGE > 0.5, recommended by Moriasi et al. (2007), was set as the threshold value for satisfactory simulations when running the auto-calibration program. The PBIAS statistics (Gupta al., 1999), which compute the average tendency of simulated variables to be larger or smaller than observed variables, were used to provide a further perspective on the model performance. With an optimum value of 0, the PBIAS values were considered satisfactory between ±0.25 for streamflow and intermediate processes, as applied in other hydrologic simulations (Hoang et al., 2017; Evenson et al., 2018b; Yeo et al., 2019).

4.2.4 Model output and hydrologic analysis

The output of SWAT-MPS is organized in three spatial scales, including the single pond, single MPS, and entire catchment. Average daily water balances for the pond portion in each of the pond HRU were first summarized across the simulation period. Average daily streamflow, baseflow and quickflow were then collected for each MPS and summarized seasonally. According to the result direction and magnitude of pond overflow, an MPS was identified as a collection of ponds with cascading fill-spill relationships, regardless of whether such connections were serial or parallel in structure, and perennial or intermittent in continuity. Meanwhile, quickflow was defined as the quantity of flow occurring immediately after the onset of a rainfall event, while baseflow was the quantity of flow after the conclusion of a rainfall event (Hornberger et al., 2014). These two flow components were calculated via SWAT Bflow (Arnold et al., 2012), which implements the baseflow separation methods in Arnold and Allen (1999) and takes daily streamflow as input. Last, average daily water balance was summarized seasonally at the catchment scale, while extreme weather events, including floods and droughts, were screened to assess the aggregated hydrologic regulation of all MPSs.

Two contrasting land use scenarios, including the BASELINE and ALL-PONDS-LOST, were simulated to evaluate the relative change in streamflow of identified MPSs and catchment water balance. The BASELINE scenario employed the present distribution and connections of MPSs, while in the ALL-PONDS-LOST scenario, MPSs were replaced with the surrounding agricultural land, owing to the pervasive and ever-increasing degradation threats (Hill et al., 2018; Chen et al., 2019). The DEM preprocessing tool in ArcSWAT was employed to fill the surface depressions, so that roughly the same subcatchments and HRUs were depicted as the BASELINE scenario.

Redundancy analysis (RDA) was performed using Canoco to examine the influences of pond characteristics, including water storage, connectivity, and bidirectional exchange with shallow groundwater, on the streamflow variations. As a constrained linear ordination technique, RDA can condense the information between two sets of variables and reveal the strength and direction of the correlation between them. Note that before RDA, data normalization, detrended correspondence analysis, and a Monte Carlo permutation test (499 permutations) were successively performed, while the resulting biplots demonstrate the correlation of two variables via their collinearity, i.e., the two arrows with a smaller angle have a higher correlation, and vice versa (see Lepš and Šmilauer, 2003 for further details).

4.3 Results
4.3.1 Calibrated parameters and model performance

Fourteen parameters from the original SWAT, including runoff, groundwater, soil, plant, and reach processes as suggested by van Griensvan et al. (2006), were initially identified by the global sensitivity analysis (Table 4.2). To avoid over-parameterization (Yen et al., 2016; Jones et al., 2019), 7 statistically significant ($p < 0.05$) parameters (among the 14 ones) in conjunction with the main stream Manning's "n" value (CH-N2), hydraulic conductivity (CH-K2), and 3 pond representation parameters were then calibrated in the GA-based optimization. Despite that the CH-N2 and CH-K2 parameters were not statistically significant, they were included to compensate for uncertainties associated with the main stream channelization in the study catchment (Chen et al., 2018a). In general, the following cross-validation resulted in a daily KGE value of 0.61 and PBIAS value of 0.11 for streamflow simulations at the outlet, and an average daily KGE value of

Hydro-Environmental Assessment of Small Water Bodies: From Local to Global Scales

0.55 and PBIAS value of -0.13 for water-level simulations for the 9 selected ponds. The calibrated SWAT-MPS was consequently considered satisfactory to represent intra-catchment processes of MPSs, although water-level simulations were slightly poorer than those for streamflow.

The calibrated parameters were evaluated by focusing on the result

Table 4.2 Calibrated model parameters and value ranges

Category	Parameter	Min.	Max.	Calibrated values	Description (units, if applicable)
Runoff	CN2	−20	20	−5 to 9	SCS runoff curve number (%)
	OV-N	−20	20	N/A[a]	Manning's "n" value for land cover to estimate overland flow (%)
	CANMX	1	10.0	2.4	Maximum canopy storage (mm)
	SURLAG	0.05	24.0	0.5 to 2.0	Surface runoff lag time (hr)
Groundwater	ALPHA-BF	0.00	1.00	N/A	Baseflow alpha factor (shallow aquifer)
	RCHR-DP	0.00	1.00	0.24	Deep aquifer percolation fraction
	GWQMN	0	500	375	Threshold depth of water in shallow aquifer (mm)
	REVAPMN	0	200	N/A	Threshold depth of water in shallow aquifer for "revap" (mm)
	GW-DELAY	10	300	N/A	Groundwater delay (days)
	GW-REVAP	0.02	0.20	N/A	Groundwater "revap" coefficient
	SOL-AWC	0.12	0.36	0.2	Available water capacity of the soil layer (mm hr^{-1})
	SOL-K	1.6	900	3.6	Saturated hydraulic conductivity (mm hr^{-1})
Soil/Plant	ESCO	0.00	1.00	N/A	Soil evaporation compensation factor
	EPCO	0.00	1.00	N/A	Plant uptake compensation factor
Reach	CH-K2	0	50	2.2	Effective hydraulic conductivity in the main stream alluvium (mm hr^{-1})
	CH-N2	0.01	0.30	0.12	Manning's "n" value for the main stream

Continued

Category	Parameter	Min.	Max.	Calibrated values	Description (units, if applicable)
Pond	P_{pond}	0	100	0 to 86	Proportion of the total overflow that routes to ponds in lower elevations (%)
	P_{stream}	0	100	0 to 42	Proportion of the total overflow that routes to subcatchment reach (%)
	K_{sat}	0	900	0.4 to 1.6	Saturated hydraulic conductivity of pond's substrate (mm hr^{-1})

[a] Non-significant parameters were not calibrated and shown as N/A (not applicable) here.

values to which they represent catchment hydrology and MPS dynamics. Groundwater parameters, including SOL-K, RCHR-DP, and GWQMN, and pond representation parameters, i.e., P_{pond}, P_{stream}, and K_{sat}, were underlined (Table 4.1). The SOL-K value (3.6 mm hr^{-1}) dictated a relative increase in saturated hydraulic conductivity throughout the SWAT HRU soil profiles, as the local data describes the soil profile as extending to a maximum of 2,100 mm below the surface (Geological Bureau of Jiangsu Province, 1980). The RCHR-DP value (0.24) dictated that 76% of water moving below the soil profile and acting as groundwater recharge within an HRU entered the shallow aquifer, while the GWQMN value (375 mm) dictated a relatively low threshold depth for groundwater leaving from the shallow aquifer to the subcatchment reach. The optimized value range of P_{pond} and P_{stream} (0-86% and 0-42%, respectively) dictated a relatively large proportion of pond water storage that overflowed to downgradient ponds and streams, while the K_{sat} values (0.4-1.6 mm hr^{-1}) dictated a far lower seepage rate of ponds' substrate than the surrounding soil profiles. These parameters, in concert, portrayed a system in which both lateral and vertical water movements exist, while the magnitude of pond spillage was prominent.

Both statistical criteria and visual comparison were inspected to verify the model performance. The daily streamflow variations were generally replicated at the catchment outlet, and most of the lower values (particularly < 0.1 m^3 s^{-1})

were successfully captured (Fig. 4.5a). However, streamflow simulations tended to be overestimated during heavy rainfall events (e.g., July 10, 2017, June 5, 2018 and August 17, 2018), presumably due to the neglect of water storage in tributary ditches and streams. With slightly larger but acceptable deviations, the 9 selected ponds were divided into 3 groups according to their elevations. The average PBIAS values for the pond group between 90-105 m (Fig. 4.5b, 4.5c, 4.5d), 75-90 m (Fig. 4.5e, 4.5f, 4.5g), and 60-75 m (Fig. 4.5h, 4.5i, 4.5j) were 0.07, -0.18, and -0.27, respectively, indicating that water-level simulation errors and associated uncertainties of fill-spill processes gradually increased from upstream to downstream. Moreover, ponds in all the 3 groups were generally wetter (i.e., higher water levels and larger submerged areas) in the simulation during rainfall periods and drier during rainless periods, which was temporally like streamflow simulations.

4.3.2 Pond water balance and MPS identification

The average daily water balance was evaluated to assess the hydrologic component of each pond (Table 4.3). Across the two-year simulation period,

Fig. 4.5 Model performance of streamflow at catchment outlet (a) and water-level at 9 selected ponds (b-j). These ponds were grouped in 3 rows according to their elevation, whose relative values to all ponds in the catchment were shown in the white-black strips (white for the highest and black for the lowest)

all 30 ponds had water storage indicated by average daily water depths within the BASELINE scenario, although a few ones (i.e., No. 4, No. 16, and No. 21) dried up for short periods. Water could enter the ponds via direct precipitation, surface inflow from drainage HRUs, spillage from upgradient ponds, and groundwater discharge. Results demonstrated that direct precipitation was the dominant means of water input for 60% of ponds, while surface inflow and spillage served as the main water source for another 30% and 10% of ponds, respectively. Water exited the ponds via ET, spillage, and groundwater recharge.

Results demonstrate that ET is the dominant means of water exiting for 73% of ponds, followed by spillage, which agreed with water balance assessments of ponds and reservoirs in humid subtropical climates (Chen et al., 2017; Tan et al., 2018). The average daily total amount of these two means for all ponds were similar (~ 109 and 93 mm d^{-1}, respectively), indicating that most of the spillage was provided by a few ponds. Moreover, the local bidirectional exchange served as groundwater recharge for 67% of ponds and discharge for the rest, while their average daily total amount was approximately proportional (~ 23 and 8 mm d^{-1}, respectively). Last, standard deviations of pond average daily water balance components were generally less than that of precipitation (Table 4.3), indicating that the ground response was smaller than precipitation variation. In viewing the meteorological data analysis in Section 4.2.1, simulated hydrologic patterns on the average daily values here and below are credible, even if extending to 5 years (slight and proportional amplification or reduction is presumably expected).

Table 4.3 Average daily values (and standard deviation) for the water balance of each pond across the simulation period

Pond No.	Area (10^3 m^2)	Average daily (standard deviation) (mm)						
		Surface inflow	Precipitation	ET	Local exchange [a]	Receive spillage	Provide spillage	Average depth [b]
1	15.01	7.86 (14.63)	3.40 (13.08)	3.68 (7.66)	1.19 (3.51)	1.76 (3.21)	10.52 (24.13)	4.03 (3.41)
2	3.84	2.65 (6.35)	3.40 (13.08)	4.91 (9.14)	0.71 (1.96)	0.00 (0.00)	1.84 (4.56)	0.68 (1.43)
3	1.49	1.82 (3.37)	3.40 (13.08)	3.45 (7.33)	1.23 (3.72)	0.00 (0.00)	2.98 (7.23)	0.71 (0.93)
4	2.08	1.31 (4.87)	3.40 (13.08)	3.36 (6.69)	0.93 (1.41)	0.46 (1.09)	2.71 (5.82)	0.32 (2.14)
5	2.78	1.62 (4.68)	3.40 (13.08)	3.79 (6.10)	−1.05 (3.12)	1.76 (4.89)	1.94 (4.81)	0.79 (1.06)
6	0.65	0.17 (0.41)	3.40 (13.08)	4.48 (6.15)	0.41 (0.97)	0.52 (2.07)	0.00 (0.00)	0.59 (1.17)

Pond No.	Area (10³ m²)	Average daily (standard deviation) (mm)						
		Surface inflow	Precipitation	ET	Local exchange [a]	Receive spillage	Provide spillage	Average depth [b]
7	0.31	0.57 (2.65)	3.40 (13.08)	3.80 (5.94)	−0.33 (1.21)	0.94 (3.19)	0.75 (2.13)	0.52 (1.08)
8	3.34	3.85 (6.68)	3.40 (13.08)	3.15 (8.41)	0.56 (2.07)	0.00 (0.00)	4.37 (9.15)	1.27 (2.48)
9	0.45	0.69 (2.67)	3.40 (13.08)	3.48 (5.82)	−0.45 (1.83)	2.61 (6.21)	2.74 (6.32)	0.71 (1.05)
10	0.38	0.22 (1.45)	3.40 (13.08)	2.89 (5.96)	−0.39 (0.89)	2.65 (5.39)	1.65 (4.13)	0.58 (3.06)
11	0.88	0.77 (2.96)	3.40 (13.08)	3.41 (6.07)	−0.45 (0.91)	0.00 (0.00)	0.30 (1.21)	0.94 (2.09)
12	0.46	0.15 (1.37)	3.40 (13.08)	3.74 (5.88)	−0.54 (1.31)	1.34 (4.67)	0.60 (1.72)	0.41 (2.05)
13	5.92	7.50 (11.63)	3.40 (13.08)	4.69 (8.97)	−2.29 (6.22)	0.67 (2.57)	4.58 (10.82)	1.07 (10.32)
14	4.47	7.14 (19.68)	3.40 (13.08)	4.19 (9.15)	0.87 (2.30)	0.00 (0.00)	7.18 (20.35)	2.16 (4.31)
15	6.95	1.72 (4.67)	3.40 (13.08)	4.70 (7.84)	−2.84 (6.37)	5.48 (14.72)	3.03 (8.29)	1.52 (6.16)
16	0.86	0.13 (2.53)	3.40 (13.08)	2.93 (6.36)	−1.06 (3.04)	0.47 (1.32)	0.00 (0.00)	0.33 (3.78)
17	2.86	3.86 (8.97)	3.40 (13.08)	3.66 (6.76)	−1.26 (3.21)	0.00 (0.00)	2.33 (6.78)	1.45 (3.11)
18	6.82	5.30 (16.78)	3.40 (13.08)	3.27 (7.75)	−2.48 (7.12)	0.86 (3.25)	3.80 (7.12)	1.62 (1.16)
19	0.87	0.16 (3.15)	3.40 (13.08)	3.22 (6.04)	−0.30 (0.89)	0.00 (0.00)	0.05 (0.33)	0.51 (1.87)
20	1.68	0.22 (5.07)	3.40 (13.08)	3.16 (5.90)	−0.91 (2.47)	0.45 (1.43)	0.00 (0.00)	0.44 (2.14)
21	0.94	2.64 (9.68)	3.40 (13.08)	3.54 (6.32)	0.66 (1.88)	0.00 (0.00)	3.13 (7.27)	0.20 (17.53)
22	0.59	0.09 (2.45)	3.40 (13.08)	2.98 (6.41)	−0.64 (1.58)	1.77 (4.32)	1.63 (3.62)	0.98 (3.52)
23	1.42	0.36 (4.14)	3.40 (13.08)	3.20 (6.10)	−0.84 (1.86)	0.27 (0.62)	0.00 (0.00)	1.71 (1.35)

Continued

Pond No.	Area (10³ m²)	Average daily (standard deviation) (mm)						
		Surface inflow	Precipitation	ET	Local exchange [a]	Receive spillage	Provide spillage	Average depth [b]
24	1.78	3.82 (14.81)	3.40 (13.08)	3.32 (6.91)	0.63 (1.64)	0.00 (0.00)	4.51 (9.21)	1.07 (1.26)
25	0.48	0.04 (4.38)	3.40 (13.08)	3.11 (5.94)	−0.25 (0.74)	0.00 (0.00)	0.06 (0.00)	1.46 (3.14)
26	0.93	0.59 (1.79)	3.40 (13.08)	3.69 (5.80)	−0.95 (1.91)	7.54 (19.39)	6.85 (11.29)	0.84 (2.17)
27	0.76	5.98 (13.37)	3.40 (13.08)	3.77 (6.67)	−1.07 (2.35)	6.17 (17.21)	10.68 (23.14)	0.61 (1.43)
28	1.30	6.13 (13.39)	3.40 (13.08)	3.72 (6.31)	0.68 (1.41)	0.00 (0.00)	6.45 (11.23)	1.13 (3.61)
29	5.75	7.98 (17.25)	3.40 (13.08)	4.17 (8.63)	−2.89 (7.17)	2.87 (6.62)	7.14 (12.39)	0.82 (0.66)
30	0.90	0.48 (4.28)	3.40 (13.08)	3.86 (7.12)	−1.58 (4.21)	3.07 (8.14)	1.48 (4.37)	0.42 (1.32)

[a] Local exchange is bidirectional, containing groundwater recharge (+, not written by default) and discharge (-).
[b] The unit of average depth is m.

According to the average daily magnitude and direction of pond spillage, 9 MPSs were identified in the catchment (Fig. 4.6a). The majority of these MPSs consisted of several ponds, where the cascading fill-spill relationship existed between pond-to-pond and pond-to-stream. A small number of MPSs (No. 2 and 4 in Fig. 4.6a) consisted of only a single pond but were also numbered for consistency. In each MPS, ponds were series-connected, e.g., No. 5, 7 and 8, or series-parallel hybrid connected, e.g., No. 1, 3, 6 and 9, with spillage to the main stream at the end or in the middle of the structure. Meanwhile, MPS drainage areas were delineated by merging the drainage area of each pond in the structure (Fig. 4.6b). The total drainage area of all MPSs took 76% of the catchment area, which partially demonstrated that pond overflow accounted for a large proportion of streamflow at the outlet. Furthermore, the drainage area of MPS No. 7, 8 and 9 was spatially separated, despite their internal spillage connections. This extraordinary finding proved the existence of fill-spill relationships in non-adjacent subcatchments, which did not follow the subcatchment and stream delineation from the 5-m resolution DEM data, but can be revealed by our rule-based knowledge integration (i.e., higher possibility of spillage connection for ponds in subcatchments with a closer topological relationship, and vice versa) and associated GA-based parameter optimization.

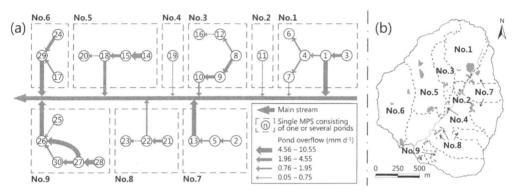

Fig. 4.6 Identified MPSs (a) and their drainage areas (b). Pond IDs from 1 to 30 in the circles in (a) are consistent with those in Fig. 4.1, while the magnitude of overflow between pond-to-pond and pond-to-stream was derived from the average daily values in Table 4.2.

4.3.3 MPS' regulation effect on streamflow, baseflow and quickflow

Variations in average daily streamflow, baseflow and quickflow of the ALL-PONDS-LOST scenario relative to the BASELINE scenario were evaluated to assess the hydrologic regulation of each MPS. The overall findings across the two-year simulation period indicated that filling MPSs decreased baseflow (~-5 to -38%), increased quickflow (~ 9 to 19%), and produced variations in the sign and magnitude of streamflow for different MPS subcatchments (see the "overall" columns in Table 4.4). Hence, the MPSs were organized into 3 control groups, according to their streamflow variations. These groups were named by the typical land use types of the MPS subcatchment, and differed significantly in the average slope: MPS No. 1, 6, 7 and 8 as the forest group (woods at 79%, agricultural land at 17%, and average slope at 16°), MPS No. 3, 5 and 9 as the farm group (agricultural land at 40%, woods at 51%, and average slope at 11°), and MPS No. 2 and 4 as the village group (rural residential areas at 39%, agricultural land at 37%, and average slope at 3°). In the subcatchments of the forest group, the average daily streamflow attenuated slightly by < 4%, relative to the BASELINE simulations. Meanwhile, streamflow in the subcatchments of the farm group witnessed a marginal increase of < 2%. The subcatchments of the village group, however, presented the notably contrasting daily streamflow of an averaged attenuation ~ 10% within the ALL-PONDS-LOST scenario.

Table 4.4 Average daily simulated streamflow of each subcatchment with MPS in the BASELINE scenario, separation of streamflow into baseflow and quickflow components, and relative percentage change for the three hydrologic components under the ALL-PONDS-LOST conditions during the simulation period

MPS No. [a]	Type	Average daily in the BASELINE scenario (10^{-3} m^3 s^{-1})					Relative change in the ALL-PONDS-LOST scenario (%)				
		Over-all	Spring [b]	Sum-mer [b]	Au-tumn [b]	Win-ter [b]	Over-all	Spring [b]	Sum-mer [b]	Au-tumn [b]	Win-ter [b]
1	Streamflow	9.20	9.23	13.00	7.36	7.19	−2.81	1.18	−8.12	−3.80	2.73
	Baseflow	5.42	5.35	7.02	4.57	4.74	−14.42	−3.15	−25.65	−20.21	−4.93
	Quickflow	3.77	3.88	5.98	2.80	2.44	13.87	11.56	16.89	13.27	10.86
2	Streamflow	0.41	0.41	0.58	0.33	0.32	−14.19	−11.85	−17.40	−15.68	−13.13
	Baseflow	0.22	0.21	0.29	0.18	0.18	−37.69	−27.17	−45.97	−34.17	−39.80
	Quickflow	0.19	0.20	0.29	0.15	0.13	12.16	9.12	13.74	14.54	9.10
3	Streamflow	2.67	2.67	3.77	2.13	2.08	1.62	−1.08	3.75	2.49	−0.27
	Baseflow	1.44	1.50	1.88	1.24	1.13	−7.68	−2.27	−3.10	−15.80	−18.70
	Quickflow	1.23	1.18	1.88	0.90	0.96	12.48	9.74	12.26	14.65	15.08
4	Streamflow	0.14	0.14	0.20	0.11	0.11	−6.01	−5.68	−7.51	−4.84	−6.87
	Baseflow	0.07	0.07	0.10	0.06	0.05	−30.05	−21.68	−39.10	−19.28	−41.27
	Quickflow	0.07	0.07	0.10	0.05	0.06	18.73	20.14	17.50	18.91	16.65
5	Streamflow	11.61	11.65	16.41	9.29	9.07	0.08	−1.84	1.68	0.71	−1.17
	Baseflow	6.84	7.22	9.85	5.02	5.26	−8.72	0.23	−3.98	−24.68	−14.90
	Quickflow	4.77	4.43	6.57	4.28	3.81	12.69	11.65	14.58	10.04	13.65
6	Streamflow	10.07	10.10	14.23	8.06	7.87	−2.27	4.26	−6.16	−8.93	3.51
	Baseflow	6.21	6.26	8.54	5.16	4.88	−10.58	−1.54	−18.76	−15.38	−2.64
	Quickflow	3.86	3.84	5.69	2.90	2.99	11.12	8.19	12.32	13.23	10.04
7	Streamflow	6.54	6.56	9.25	5.24	5.11	−4.20	2.51	−10.12	−7.19	1.18
	Baseflow	4.04	4.07	5.37	3.25	3.48	−14.27	−1.65	−22.27	−28.01	−4.41
	Quickflow	2.50	2.49	3.89	1.99	1.64	12.05	10.51	15.62	10.62	7.94

MPS No. [a]	Type	Average daily in the BASELINE scenario (10^{-3} m³ s⁻¹)					Relative change in the ALL-PONDS-LOST scenario (%)				
		Over-all	Spring [b]	Sum-mer [b]	Au-tumn [b]	Win-ter [b]	Over-all	Spring [b]	Sum-mer [b]	Au-tumn [b]	Win-ter [b]
8	Streamflow	2.01	2.01	2.84	1.61	1.57	−0.76	0.84	−2.95	−2.12	1.35
	Baseflow	1.17	1.21	1.59	0.93	0.97	−9.71	0.37	−19.32	−9.87	−6.31
	Quickflow	0.83	0.81	1.25	0.67	0.60	11.90	10.68	13.75	10.00	10.68
9	Streamflow	4.86	4.88	6.87	3.89	3.80	0.70	−2.13	1.27	1.39	2.14
	Baseflow	2.94	2.93	3.99	2.41	2.43	−4.92	−2.68	−0.51	−13.80	−6.71
	Quickflow	1.92	1.95	2.89	1.48	1.37	9.31	6.18	10.08	12.15	8.68

[a] Subcatchment of each numbered MPS in Fig. 4.6. Note: MPS No. 2 and 4 had a single pond but numbered for consistency.
[b] Aggregated by the same season across the simulation period.

Seasonal variations in streamflow, baseflow and quickflow of the ALL-PONDS-LOST scenario were then evaluated, and it was found that MPSs had seasonally-independent regulation effect on the average daily streamflow and baseflow, while quickflow regulation was comparatively consistent across seasons (see the season columns in Table 4.3). Relative to the BASELINE simulations, for example, the subcatchments of the forest group produced attenuated average daily streamflow during summer and autumn (~-2 to-10%) but increased the average daily streamflow during spring and winter (~ 1 to 4%). The filling of MPS in these subcatchments markedly lowered the average daily baseflow within summer and autumn (~-10 to-28%) but presented moderate attenuation in spring and winter (>-6%). Average daily quickflow in the 4 subcatchments increased across seasons by ~ 8 to 17% relative to the BASELINE scenario.

In the ALL-PONDS-LOST scenario, the subcatchments of the farm group had an opposite trend in average daily streamflow relative to the above subcatchments, but similar seasonally-independent regulation effects on streamflow and baseflow. In the subcatchments of MPS No. 3 and 5, average daily streamflow increased in summer and autumn (< 4%), while decreasing in spring and winter (>-2%). In the subcatchment of MPS No. 9, similar variations occurred in spring, summer and autumn (~-2, 1 and 1%, respectively), but a reverse trend was found in winter (~ 2%). Further, average daily baseflow simulations within the three subcatchments suggested marked attenuations in autumn and winter (~-7 to-25%), yet comparatively smaller attenuations in spring and summer (>-3%), despite a marginal increase (0.23%) in spring for the subcatchment of MPS No. 5. Again, average daily quickflow in the three subcatchments increased across seasons by ~ 6 to 15%.

The subcatchment of the village group also had contrasting simulations. Relative to the BASELINE simulations, the subcatchment of MPS No. 2 witnessed a moderate attenuation in the average daily streamflow (~-12 to-17%), despite that a smaller decline was found in the subcatchment of MPS No. 4 (~-5 to-8%). Attenuation in the average daily baseflow was similar and drastic (~-19 to-46%) in the two subcatchments but increase in the average daily quickflow was observed to be greater across seasons in MPS No. 4 (~ 17 to 20%) than that in MPS No. 2 (~ 9 to 15%).

4.3.4 Catchment water balance and extreme hydrographs

To assess the aggregated regulation effects of all MPSs, simulated catchment water balance of the ALL-PONDS-LOST scenario was compared with that of the BASELINE scenario (Table 4.5). Results indicated that potential evapotranspiration (PET) values decreased slightly after filling the MPSs (area proportion at ~ 5%) and replacing with agricultural land, especially in spring and summer (~-0.6%). Such attenuation was more notable in ET, which was ~-4% across the years, and had two low values in summer (~-5.5%) and autumn (-5%). Similarly, the soil water content of the catchment declined across seasons at ~-7 to-4%. Without the pond water storage and intra-catchment fill-spill processes, surface flow at the outlet significantly increased by ~ 13% overall and peaked in summer by 15%. Groundwater flow, however, declined across seasons (~-20 to-11%), and reached the minimum value in winter. The total water yield at the catchment outlet decreased by 3% overall, which almost levelled off during spring and summer (~ -0.4%) yet decreased markedly during autumn and winter (~-6%).

Table 4.5 Average daily simulations of catchment water balance, and relative percentage changes of the ALL-PONDS-LOST to BASELINE scenario

Scenarios	Hydrologic components†	Average daily (mm)					Relative change (%)				
		Overall	Spring	Summer	Autumn	Winter	Overall	Spring	Summer	Autumn	Winter
BASELINE	Precipitation	3.40	3.04	6.80	2.91	1.78	N/A [b]	N/A	N/A	N/A	N/A
	PET	3.02	3.35	4.86	2.63	1.22	N/A	N/A	N/A	N/A	N/A
	ET	1.57	2.02	2.56	1.19	0.51	N/A	N/A	N/A	N/A	N/A
	Soil water content [a]	146.47	140.69	153.42	148.31	143.47	N/A	N/A	N/A	N/A	N/A
	Surface flow	1.04	1.05	1.47	0.84	0.82	N/A	N/A	N/A	N/A	N/A
	Groundwater flow	1.29	1.14	1.65	1.38	0.97	N/A	N/A	N/A	N/A	N/A
	Total water yield	2.33	2.19	3.12	2.22	1.79	N/A	N/A	N/A	N/A	N/A
ALL-PONDS-LOST	Precipitation	3.40	3.04	6.80	2.91	1.78	0.00	0.00	0.00	0.00	0.00
	PET	3.01	3.33	4.83	2.62	1.22	−0.39	−0.60	−0.62	−0.38	0.00
	ET	1.50	1.96	2.42	1.13	0.50	−4.30	−2.97	−5.47	−5.04	−1.96
	Soil water content	139.22	135.25	142.84	143.05	135.73	−4.95	−3.87	−6.90	−3.55	−5.39
	Surface flow	1.18	1.16	1.69	0.94	0.91	12.98	10.48	14.97	11.90	10.98
	Groundwater flow	1.10	1.02	1.44	1.14	0.78	−14.79	−10.53	−12.73	−17.39	−19.59
	Total water yield	2.27	2.18	3.13	2.08	1.69	−2.58	−0.46	0.32	−6.31	−5.59

[a] Under the average total depth of the two soil layers in the study catchment.
[b] Relative change (%) is N/A (not applicable) for the BASELINE simulations themselves.

Simulated hydrographs of the BASELINE and ALL-PONDS-LOST scenario were further inspected to assess the MPS aggregated regulation effect on hydrologic extremes. The screened events included short-term and heavy rainfall (Fig. 4.7a), smaller but intermittent rainfall (Fig. 4.7b), and two severe droughts for the peripheral regions, each lasting more than a month (Fig. 4.7c and 4.7d). Results suggested that flood resilience of MPSs was more pronounced during heavier rainfall, as the simulated streamflow in the BASELINE scenario exhibited ~ 0.4 m^3 s^{-1} (28%) less than that of the ALL-PONDS-LOST scenario during the largest rainy day in Fig. 4.7a, while such difference averaged at 0.05 m^3 s^{-1} (20%) during the three rainfall peaks in Fig. 4.7b. The hydrographs of the two scenarios intersected as time went on after the rainfall events. When the streamflow returned to baseflow in Fig. 4.7c and during the dry days in Fig. 4.7d, the average daily values in the BASELINE scenario increased by ~ 26% (0.01 m^3 s^{-1}) relative to that in the ALL-PONDS-LOST scenario, indicating that the presence of MPSs resulted in consistently higher simulated baseflow compared to that in their absence.

Fig. 4.7 Simulated hydrographs under BASELINE and ALL-PONDS-LOST conditions during 4 hydrologic extremes. Plot (a) shows a heavy rainfall event totaled 167.8 mm during 16 to 20 August 2018. Plot (b) shows a smaller but intermittent rainfall event totaled 77.5 mm from 17 May to 10 June 2018. Plot (c) and (d) present two droughts in 2017, in which the former followed a moderate rainfall event (37.9 mm) on 10 June, while the latter lasted from October to December.

4.4 Discussion
4.4.1 Multi-scale hydrologic regulation of MPSs

Small, scattered wetlands, such as vernal pools, prairie potholes, and cypress domes, provide numerous ecosystem services by altering the distribution and form of water resources locally and regionally (Verhoeven et al., 2006; Hill et al., 2018; Golden et al., 2019). Among these wetlands, MPSs, which date back over 2000 years, are often dominant on small farms in southern China, but are poorly quantified regarding their intra-catchment hydrologic processes (Yu, 2015; Chen et al., 2019; Liu et al., 2009). In this context, the hydrologic regulations of MPS were explored over numerous scales via an enhanced semi-distributed and semi-physically based SWAT model, which includes the following advancements: (1) parameterizing the cascading fill-spill relationships of pond-to-pond and pond-to-stream in a spatially-explicit pond representation framework, (2) estimating bidirectional surface and subsurface water exchange for each pond according to the water-level gradients with surrounding drinking water wells, and (3) employing rule-based artificial intelligence for model calibration and validation to elucidate MPS structures (i.e., ponds with perennial or intermittent spillage connections).

Previous studies have concluded that wetlands attenuate peak flows at the catchment scale (Feng et al., 2013; Golden et al., 2014; Lee et al., 2018; Yeo et al., 2019). This axiom was supported by our hydrograph comparisons between the BASELINE and ALL-PONDS-LOST simulations, as peak flows reduced more than 20% during the heavy rainfall events in Fig. 4.7a and 4.7b. In addition to the intermediate pond water storage, flood resilience was mainly due to the cascading fill-spill processes of identified MPSs, which reinforced the lag effect summarized by Rains et al., 2016. When rainfall occurred within SWAT-MPS, water first entered the ponds via direct precipitation and drainage (i.e., surface runoff), while overflow to the main stream only occurred when all upgradient ponds in the MPS structure were filled. Moreover, the lag effect can also increase flood residence times (Rahman et al., 2016; Evenson et al., 2018a; Ameli and Creed, 2019), but this was not apparent in our small catchment (Fig. 4.7a and 4.7b). Instead, the hydrographs of the two scenarios intersected when streamflow returned to baseflow several days after rainfall peaks in Fig. 4.7c and 4.7d, indicating that MPSs increased soil water content and groundwater flow via the relatively slow infiltration process, and thus raised baseflow at the catchment outlet. These inferences were consistent with the simulated catchment water balance in Table 4.4, as decreased percentage of soil water content and groundwater flow were revealed after filling the MPSs. Meanwhile, pond spillage was equally important, according to the average daily baseflow and quickflow in Table 4.4 and surface and groundwater flow in Table 4.5. Hence, comparing with low relief and geologically young landscapes in the US Southeastern Coastal Plain (Evenson et al., 2018b, Lee et al., 2018, and Yeo et al., 2019), and polder areas in Netherlands (Yu et al., 2018) and Taihu Lake Basin (Yan et al., 2018), where groundwater plays a dominant role in the hydrologic circle, and situations of MPSs in low mountains and hills are different.

At the MPS scale, the effect of baseflow support and quickflow attenuation was consistent but varied in degree across different MPSs and seasons (Table 4.3). Such differences can be attributed to pond intrinsic characteristics (e.g., water storage, connectivity, and bidirectional exchange with shallow groundwater) and extrinsic factors of subcatchment (e.g., slope, land use types, and soil attributes) and precipitation (e.g., intensity, duration, and concentration) (Golden et al., 2014; Yasarer et al., 2018; Jones et al., 2019). Under the same rainfall conditions but disparate subcatchment factors, RDA was performed to discuss the influences of pond characteristics on baseflow and quickflow variations in the 3 MPS groups (forest, farm, and village narrated in Section 4.3.3) (Fig. 4.8). Results indicated apparent influences of pond water storage for baseflow support, and the correlation ranked: forest > farm > village group in our study catchment. We attributed the strongest linkage to the low human disturbance and the large drainage area associated with pond water storage in the natural woodlands, and the weakest linkage to the impervious land surface and low positions of the village, where baseflow may primarily come from upgradient subcatchments. Moreover, the influence of pond water storage was stronger than spillage in the forest group, but slightly weaker than that in the farm group. Such difference was consistent with the complexity of surface water connections in the two groups, where the pervasive and interlaced ditch network in the farmland can contribute to a more spatially

Fig. 4.8 Redundancy analysis biplots of the MPS forest (a), farm (b), and village (c) group, showing correlations between pond variables (red dotted lines) and relative changes in baseflow and quickflow (blue solid lines) under the ALL-PONDS-LOST conditions. The variables include aggregated water storage (Storage), magnitude of spillage received and provided (Spillage), and bidirectional exchange with shallow groundwater (Recharge and Discharge) of all ponds in each MPS subcatchment and were reckoned by the average daily values in Table 4.2.

enough baseflow support (Liu et al., 2009; Tan et al., 2018).

For quickflow attenuation, the influence of pond spillage was stronger in the forest group than that in the village group, while the correlation was found negative in the farm group. We attributed the relative stronger correlation to the concurrent flow path of pond-to-pond and pond-to-stream in the woodlands, while the negative correlation in the farmland to (1) intentional blockage of the ditch network during rainfall to increase pond water storage and reduce farmland soil erosion (Chen et al., 2019), and (2) the perennial saturated (high water-level) condition of some downgradient, aquaculture ponds, e.g., No. 8 and No. 18, that may adversely contribute to a higher flood peak (Mitsuo et al., 2014; Wu et al., 2020). Due to the indirect regulation effect that determined by pond water-level, the influence of bidirectional water exchange was rather limited, comparing to water storage and spillage (Fig. 4.8). Even so, the correlation of groundwater recharge and quickflow attenuation was demonstrated across all MPS groups, owing to the increased travel time and hydrologic path (Golden et al., 2014). Our findings, including (1) the ranked level of pond water storage for baseflow support across different land use groups, and (2) the inversed correlation of pond spillage to baseflow and quickflow variation in the farmland, presented site-specific answers to the long-standing question about how intrinsic characteristics and extrinsic land use types jointly affect the hydrologic regulation of wetlands (Rains et al., 2016;

Golden et al., 2019; Jones et al., 2019). The total explained variance of the two axes in RDA (< 40%), however, revealed other potential but stronger correlation factors, such as precipitation intensity and duration, which develop further research interests applicable to larger watershed and river basins, but was beyond the scope of the current study.

According to the simulated water balance at the pond scale, 27 out of the 30 ponds had water storage throughout the simulation period. As small, scattered wetlands are reported to dry out in other landscapes (Williams and Fryirs, 2016; Poschlod and Braun-Reichert, 2017; Evenson et al., 2018b), we attributed the submerged conditions to the humid climate that produces direct precipitation and drainage to single ponds, and more importantly, the cascading fill-spill processes that keep balanced water storage across all ponds in the MPS structure. Moreover, 8 ponds (~ 27% in total) of the 7 structures simultaneously provided spillage to the main stream and received spillage from upgradient ponds (Fig. 4.6). The average daily overflow provided and received by these ponds totaled 1.37 and 0.72 mm (converted to runoff depth and excluded evapotranspiration losses), which took ~ 59 and 31% of the catchment water yield in Table 4.4, and again portrayed a nonnegligible role of pond fill-spill processes to the catchment hydrology. Different from analytical scenarios that partition the wetlands according to their distance to streams (Evenson et al., 2018a; Golden et al., 2019), or their spatial distribution in the catchment (e.g., upstream, midstream and downstream) (Hansen et al., 2018; Lee et al., 2018), our study elucidated the actual connecting structures and quantified the hydrologic components of each pond, MPS, and entire catchment, which support and refine the emerging findings that wetland connectivity and hydrologic fluxes are crucial to their catchment functions, in addition to the number, area, and distribution (Cohen et al., 2016; Lane et al., 2018; Ameli and Creed, 2019).

4.4.2 Implications for management

With an estimated global loss of about 35% since the 1970s, wetlands are one of the most threatened ecosystems, and are central to meeting many of the United Nations' 17 Sustainable Development Goals (Gardner and Finlayson, 2018). Although a no-net-loss target for wetlands has been set in many nations, most conservation policies ignore small, scattered waters, resulting in hidden loss of wetland functions behind the reported "paper offsets" (UN-Water, 2018; Xu et al., 2019). Following our findings that both pond water storage and spillage connections are important to their hydrologic regulation, conservation efforts do not necessarily require expensive engineering measures to create new wetlands but need to maintain and optimize the perennial or intermittent surface water connections between the existing ones. A relevant practice is pond and ditch dredging to reduce siltation during agricultural slack seasons in China and Japan, which is traditional, cost-efficient, but increasingly ignored owing to the limited short-term economic benefit (Yin et al., 2006; Mitsuo et al., 2014; Yu et al., 2015). Hence, we propose encouragement and revitalization of similar traditional measures to ensure connections and flowing waters for wetland functions—not limited to MPSs in the Chinese context, but similar small, scattered wetlands elsewhere.

Tested in low mountains and hills under a humid subtropical climate, our work confirmed that incorporating intra-catchment observations benefits wetland and catchment modelling (Yen et al., 2016; Golden et al., 2017; Lane et al., 2018). To acquire generalized knowledge on MPSs' hydrologic regulation and guide pond management in larger watersheds and river basins, we call on several measures to improve the monitoring system and possible data scarcities. First, to place an emphasis on pond mapping, as these small, scattered waters are usually disregarded in wetland inventories (Tiner et al., 2015; Wu, 2018). Concerted efforts from government sectors, including water resources, agriculture, and rural development, are preferred, since our interviews with the local farmers indicated that MPS structures are easily damaged by agricultural activities, and even filled during land consolidation (Long, 2014). Second, to encourage long-term monitoring on pond dynamics, such as water-level variations, despite the reported deficiencies since the 2000s (Whigham and Jordan, 2003; Hill et al., 2018; Zhang et al., 2020). Facilitated by various smart devices and wireless networks, this work can be voluntarily or encouraged through incentives by locals after holding workshops to reinforce their responsibility as pond managers, as has been tested in farmer-participatory rural reform in our study catchment (Chen et al., 2018). Furthermore, remote sensing techniques, including light detection and ranging (LiDAR), synthetic aperture radar (SAR), and multispectral scanning (MSS), have inherent abilities to characterize these waters (Chen et al., 2019; Wu, 2019), and can cross-validate and reduce uncertainties of the simulated fill-spill processes.

Reported roles of small, scattered wetlands go beyond hydrologic regulation and include nutrient (nitrogen and phosphorus) buffers to safeguard downstream rivers and lakes against eutrophication (Verhoeven et al., 2006; Zhang et al., 2012; Xia et al., 2018), as well as freshwater habitats and refuges for endangered plants, invertebrates and amphibians (Mitsuo et al., 2014; Williams and Fryirs, 2016; Poschlod and Braun-Reichert, 2017). Partly driven by the submerged conditions and perennial or intermittent water connections, the environmental and ecological concerns have become a well-studied topic in large aquatic systems, such as dam-regulated rivers (Grantham et al., 2014), but are seldom assessed in conjunction with the hydrologic regulation of MPS during rural planning and pond management. Based on identified direction and magnitude of pond overflow, hydrodynamic models, such as Water Quality Analysis Simulation Program (WASP) and Environmental Fluid Dynamics Code (EFDC), have potential to reveal the flow paths within ponds and ditches, and their associated influences on the surrounding water quality and flora and fauna. Hence, the integration of these models was recommended to make wider application of our detailed hydrologic assessment and identify the overlapped or interlaced hotspots of MPSs that require particular conservation efforts.

4.4.3 Improved model limitations and recommendations

We present a novel contribution toward quantifying the intra-catchment processes of small, scattered wetlands compared to previous studies. However, all modelling methods necessarily have limitations. For example, the direction and magnitude of pond spillage (i.e., P_{pond} and P_{stream}) was inversely identified by streamflow and water-level variations in conjunction with the pre-defined production rules and GA-based parameter optimization. Screening from several qualified GA populations, the rule-based artificial intelligence was validated by the statistical criteria, but brought uncertainties during subcatchment delineation and rule formulation, owing to the multiple ponds within a single subcatchment (Fig. 4.3a), and various fill-spill possibilities between different subcatchments (Table 4.1). In view of the increased errors of simulated water-level from upstream to downstream narrated in Section 4.3.1, the uncertainties may be minimized by reducing the subcatchment size so that each one contains a single pond representation but may also increase the rule complexity and screening procedures to obtain qualified parameter values. To balance their respective complexity and model performance, the relationship between the pond number in each subcatchment and the rule-based fill-spill possibility requires further digestion, although the present method is effective in depicting intra-catchment processes and its SWAT-independent implementations are referential to similar catchment modelling.

Our SWAT enhancements did not necessarily signify improved model performance as measured by standard statistical criteria. To verify this point, an original SWAT with default pond representation was applied in the study catchment during the 2-year simulation period. Using the calibrated parameters of SWAT-MPS in the BASELINE scenario, the original model had daily KGE and PBIAS values of 0.73 and -0.09 for streamflow simulations, respectively. Comparing with the performance in Section 4.3.1, the two models resulted in relatively similar KGE values, while the original model tended to underestimate streamflow (PBIAS value of -0.09), and the new model tended to overestimate streamflow (PBIAS value of 0.11). Relative to the employed parameters that govern the rate of water transport in the simulated main stream (i.e., CH-N2 and CH-K2), several SWAT extensions can potentially improve the performance of streamflow simulation in the new model, such as explicit representations of riparian wetlands and tributary ditches and streams (Rahman et al., 2016), as well as water pumping from streams to ponds that usually happens before irrigation seasons to increase pond water storage (Tan et al., 2018). However, streamflow-oriented model improvements are not enough for SWAT-MPS, partly because of the equifinality phenomenon in auto-calibration, i.e., several different combinations of model parameters can result in similar predictions (Efstratiadis and Koutsoyiannis, 2010; Yen et al., 2016), and partly the ignorance of intra-catchment processes in performance evaluation, including pond spillage connections, variations in submerged area, and bidirectional exchanges with shallow groundwater.

Future work must evaluate SWAT-MPS against more diverse monitoring data, over longer simulation periods. For example, the bidirectional exchange is recognized as an important hydrologic component of pond water balance in humid agricultural regions (Golden et al., 2014; Ameli and Creed, 2017; Yeo et al., 2019). However, SWAT-MPS employed an aggregate rate of vertical and lateral seepage, owing to the parameterization of only one hydraulic conductivity value to simulate seepage from HRUs. Calibrated by water-level gradients between ponds and drinking water wells, the vertical exchange was separated and valued between -2.89 to 1.23 mm d^{-1} for the study catchment (Table 4.2), on the condition that wells are deep to the same aquifer. This areal result was approximate to SWAT simulations in peripheral catchments (Lin et al., 2013), but requires downscaling before comparing with soil seepage observations at the slope scale (Vanderlinden et al., 2012; Li et al., 2015). Moreover, other field observations, such as streamflow between some large, perennially connected ponds, chemical and isotopic tracings (Thorslund et al., 2018), and microbial communities and indicators (Mushet et al., 2019), can be used to validate the spillage connections that characterize MPSs. A combined use of these datasets over longer periods can enhance MPSs' multi-scale hydrologic assessment under various precipitation conditions, and advance our understanding of how small, scattered wetlands affect catchment functions via their structures and processes.

4.5 Conclusions

Assessing wetland hydrologic processes over numerous scales is crucial in understanding their ecosystem services especially nutrient buffering. In this study, we demonstrate a novel approach to quantify the multi-scale hydrologic regulation of multi-pond systems (MPSs), a common type of small, scattered wetland in humid agricultural regions. Our new model, dubbed SWAT-MPS, incorporated improved representations of: (1) cascading fill-spill relationships of pond-to-pond and pond-to-stream, and (2) bidirectional exchange between pond surface water and shallow groundwater. Using rule-based knowledge integration and GA-based parameter optimization, SWAT-MPS successfully replicated streamflow and pond water-level variations in the 4.8 km^2 test catchment, southern China. Water balance analysis revealed the presence of 9 series- or series-parallel connected MPSs, in which pond overflow accounted for as much as 59% of the catchment water yield. Scenario simulations revealed seasonally- and MPS-independent baseflow support and quickflow attenuation at the MPS scale, while at the catchment scale, MPSs in aggregate, reduced flood peaks by > 20% and increased baseflow by 26% in the following dry days. Redundancy analysis further discovered (1) the ranked level of pond water storage for baseflow support across different land use types: forest > farm > village, and (2) the inversed correlation of pond spillage to baseflow and quickflow variations in the farmland. Preliminary meteorological data analysis and resulting average daily values confirmed the credibility of our MPS hydrologic patterns, even if extending the 2-year simulation to longer periods.

This study underscores MPS's water storage and connectivity in its flood and drought resilience, and encourages inventories, long-term field monitoring, and integration of hydrodynamic models for integrated pond management in watersheds and river basins. Despite potential research directions, such as balancing the complexity of subcatchment and production rules to reduce pond fill-spill uncertainties, and incorporating diverse, long-term monitoring data to fully reflect pond dynamics in performance evaluation, our model enhancements and analytical results serve as a first exploration on the intra-catchment hydrologic details of MPSs in the Chinese context. They can also inform refined assessment of similar small, scattered wetlands elsewhere, where restoration efforts are required.

References

Abbaspour, K. C., 2014. "SWAT-CUP 2012: SWAT calibration and uncertainty program-A user manual." *Science and Technology* 106.

Ameli, A. A., and Creed, I. F., 2017. "Quantifying hydrologic connectivity of wetlands to surface water systems." *Hydrology and Earth System Sciences* 21(3): 1791-1808.

Ameli, A. A., and Creed, I. F., 2019. "Does wetland location matter when managing wetlands for watershed-scale flood and drought resilience." *Journal of the American Water Resources Association* 55(3): 529-542.

Arnold, J. G., and Allen, P. M., 1999. "Automated methods for estimating baseflow and ground water recharge from streamflow records." *Am. Water Resour. As.* 35(2): 411-424.

Arnold, J. G., Kiniry, J. R., Srinivasan, R., et al., 2012. Soil and water assessment tool input/output documentation version 2012. Texas Water Resources Institute.

Chen, W., He, B., Ma, J., et al, 2017. "A WebGIS-based flood control management system for small reservoirs: A case study in the lower reaches of the Yangtze River." *J. Hydroinform.* 19(2): 299-314.

Chen, W., He, B., Nover, D., et al., 2018a. "Spatiotemporal patterns and source attribution of nitrogen pollution in a typical headwater agricultural watershed in Southeastern China." *Environ. Sci. Pollut. R.* 25(3): 2756-2773.

Chen, W., He, B., Nover, D., et al., 2019. "Farm ponds in southern China: Challenges and solutions for conserving a neglected wetland ecosystem." *Sci. Total Environ.* 659: 1322-1334.

Chen, W., Nover, D., He, B., et al., 2018b. "Analyzing inundation extent in small reservoirs: A combined use of topography, bathymetry and a 3D dam model." *Measurement* 118(9): 202-213.

Chlumecký, M., Buchtele, J., Richta, K., 2017. "Application of random number generators in genetic algorithms to improve rainfall-runoff modelling." *J. Hydrol.* 553: 350-355.

Cohen, M. J., Creed, I. F., Alexander, L., et al., 2016. "Do geographically isolated wetlands influence landscape functions." *P. Natl. Acad. Sci. USA.* 113(8): 1978-1986.

Dai, L., 2019. "Implementing the water goals-The river chief mechanism in China." *In Politics and Governance in Water Pollution Prevention in China.* London: Palgrave Pivot.

Efstratiadis, A., and Koutsoyiannis, D., 2010. "One decade of multi-objective calibration approaches in hydrological modeling: A review." *Hydrol. Sci. J.* 55 (1): 58-78.

Engelmore, R. S., and Feigenbaum, E., 1993. "Expert systems and artificial intelligence." *Expert Systems* 100(2): 1-7.

Evenson, G. R., Golden, H. E., Lane, C. R., et al., 2015. "Geographically isolated wetlands and watershed hydrology: A modified model analysis." *J. Hydrol.* 529: 240-256.

Evenson, G. R., Golden, H. E., Lane, C. R., et al., 2018a. "Depressional wetlands affect watershed hydrological, biogeochemical, and ecological functions." *Ecol. Appl.* 28(4): 953-966.

Evenson, G. R., Jones, C. N., McLaughlin, D. L., et al., 2018b. "A watershed-scale model for depressional wetland-rich landscapes." *J. Hydrol. X.* 1:100002.

Feng, X. Q., Zhang, G. X., Jun, X. Y., 2013. "Simulation of hydrological processes in the Zhalong wetland within a river basin, Northeast China." *Hydrol. Earth Syst. Sci.* 17(7): 2797-2807.

Foglia, L., Mehl, S. W., Hill, M. C., et al., 2007. "Testing alternative ground water models using cross-validation and other methods." *Groundwater* 45(5): 627-641.

Fortin, F. A., Rainville, F. M. D., Gardner, M. A., et al., 2012. "DEAP: Evolutionary algorithms made easy." *J. Mach. Learn. Res.* 13: 2171-2175.

Gardner, R. C., and Finlayson, M., 2018. Global wetland outlook: State of the world's wetlands and their services to people. Secretariat of the Ramsar Convention, Gland, Switzerland.

Geological Bureau of Jiangsu Province, 1980. Regional hydrogeological survey of the People's Republic of China: Zhenjiang region, 66-76.

Goldberg, D., 1989. *Genetic algorithms in search optimization and machine learning.* Boston: Addison-Wesley.

Golden, H. E., Creed, I. F., Ali, G., et al., 2017. "Integrating geographically isolated wetlands into land management decisions." *Front. Ecol. Environ.* 15(6): 319-327.

Golden, H. E., Lane, C. R., Amatya, D. M., et al., 2014. "Hydrologic connectivity between geographically isolated wetlands and surface water systems: A review of select modeling methods." *Environ. Modell. Softw.* 53: 190-206.

Golden, H. E., Rajib, A., Lane, C. R., et al., 2019. "Non-floodplain wetlands affect watershed nutrient dynamics: A critical review. Environ." *Sci. Technol.* 53(13): 7203-7214.

Grantham, T. E., Viers, J. H., Moyle, P.B., 2014. "Systematic screening of dams for environmental flow assessment and implementation." *BioScience* 64(11): 1006-1018.

Gupta, H. V., Kling, H., Yilmaz, K. K., et al., 2009. "Decomposition of the mean squared error and NSE performance criteria: Implications for improving hydrological modelling." *J. Hydrol.* 377(1-2): 80-91.

Gupta, H. V., Sorooshian, S., Yapo, P. O., 1999. "Status of automatic calibration for hydrologic models: Comparison with multilevel expert calibration." *J. Hydrol. Eng.*4(2): 135-143.

Hansen, A.T., Dolph, C.L., Foufoula-Georgiou, E., et al., 2018. "Contribution of wetlands to nitrate removal at the watershed scale." *Nat. Geosci.* 11(2): 127-132.

Herschy, R. W., 2008. *Streamflow measurement.* Boca Raton: CRC Press.

Hill, M. J., Hassall, C., Oertli, B., et al., 2018. "New policy directions for global pond conservation." *Conserv. Lett.* 11(5): e12447.

Hoang, L., Schneiderman, E. M., Moore, K. E., et al., 2017. "Predicting saturation-excess runoff distribution with a lumped hillslope model: SWAT-HS." *Hydrol. Process.* 31(12): 2226-2243.

Hornberger, G. M., Wiberg, P. L., Raffensperger, J. P., et al., 2014. *Elements of physical hydrology.* Baltimore: The John Hopkins University Press.

Jones, C. N., Ameli, A., Neff, B. P., et al., 2019. "Modeling connectivity of non-floodplain wetlands: Insights, approaches, and recommendations." *J. Am. Water Resour. As.* 55(3): 559-577.

Kang, H., and Sridhar, V., 2017. "Combined statistical and spatially distributed hydrological model for evaluating future drought indices in Virginia." *Journal of Hydrology: Regional Studies* 12(C): 253-272.

Lane, C. R., Leibowitz, S. G., Autrey, B. C., et al., 2018. "Hydrological, physical, and chemical functions and connectivity of non-floodplain wetlands to downstream waters: A review." *J. Am. Water Resour. As.* 54(2): 346-371.

Lee, S., Yeo, I. Y., Lang, M. W., et al., 2018. "Assessing the cumulative impacts of geographically isolated wetlands on watershed hydrology using the SWAT model coupled with improved wetland modules." *J. Environ. Manage.* 223: 37-48.

Lepš, J., and Šmilauer, P., 2003. *Multivariate analysis of ecological data using CANOCO.* New York: Cambridge University Press.

Li, Q., Zhu, Q., Zheng, J., et al., 2015. "Soil moisture response to rainfall in forestland and vegetable plot in Taihu Lake Basin, China." *Chinese Geogr. Sci.* 25(4): 426-437.

Lin, B., Chen, Y., Chen, X., 2013. "A study on regional difference of hydrological parameters of SWAT model." *Journal of Natural Resource* 28(11): 1988-1999.

Liu, Y., Fu, Q., Yin, C., 2009. "Phosphorus sorption and sedimentation in a multi-pond system within a headstream agricultural watershed." *Water Qual. Res. J. Can.* 44(3): 243-252.

Long, H., 2014. "Land consolidation: An indispensable way of spatial restructuring in rural China." *J. Geogr. Sci.* 24(2): 211-225.

Longbucco, N., 2010. *Impacts of small water bodies on the hydrological response of small agricultural watersheds.* Carbondale: Southern Illinois University.

Ly, S., Charles, C., Degré, A., 2013. "Different methods for spatial interpolation of rainfall data for operational hydrology and hydrological modeling at watershed scale: A review." *Biotechnol. Agron. Soc.* 17(2): 392-406.

Mitsuo, Y., Tsunoda, H., Kozawa, G., et al., 2014. "Response of the fish assemblage structure in a small farm pond to management dredging operations." *Agr. Ecosyst. Environ.* 188: 93-96.

Moriasi, D. N., Arnold, J. G., Van Liew, M. W., et al., 2007. "Model evaluation guidelines for systematic quantification of accuracy in watershed simulations." *T. ASABE.* 50(3): 885-900.

Mushet, D. M., Alexander, L. C., Bennett, M., et al., 2019. "Differing modes of biotic connectivity within freshwater ecosystem mosaics." *J. Am. Water Resour. As.* 55(2): 307-317.

Nash, J. E., and Sutcliffe, J. V., 1970. "River flow forecasting through conceptual models part I-A discussion of principles." *J. Hydrol.* 10(3): 282-290.

Poschlod, P., and Braun-Reichert, R., 2017. "Small natural features with large ecological roles in ancient agricultural landscapes of Central Europe-History, value, status, and conservation." *Biol. Conserv.* 211: 60-68.

Rahman, M. M., Thompson, J. R., Flower, R. J., 2016. "An enhanced SWAT wetland module to quantify hydraulic interactions between riparian depressional wetlands, rivers and aquifers." *Environ. Modell. Softw.* 84: 263-289.

Rains, M. C., Leibowitz, S. G., Cohen, M. J., et al., 2016. "Geographically isolated wetlands are part of the hydrological landscape." *Hydrol. Process.* 30(1): 153-160.

Saxton, K. E., and Rawls, W., 2005. Soil water characteristics: Hydraulic properties calculator. Agricultural Research Service, United States Department of Agriculture.

Scheidegger, A. E., 1965. "The algebra of stream-order numbers." *United States Geological Survey Professional Paper* 525: 187-189.

Shafii, M., and Smedt, F. D., 2009. "Multi-objective calibration of a distributed hydrological model (WetSpa) using a genetic algorithm." *Hydrol. Earth Syst. Sci.*13(11): 2137-2149.

Sophocleous, M., 2002. "Interactions between groundwater and surface water: The state of the science." *Hydrogeol. J.* 10(1): 52-67.

Sridhar, V., Kang, H., Ali, S. A., 2019. "Human-induced alterations to land use and climate and their responses for hydrology and water management in the Mekong River Basin." *Water* 11(6): 1307.

Tan, X., Shao, D., Gu, W., 2018. "Improving water reuse in paddy field districts with cascaded on-farm ponds using hydrologic model simulations." *Water Resour. Manag.*32(5): 1849-1865.

Tarboton, D. G., 2005. "Terrain analysis using digital elevation models in hydrology." Utah State University.

Thorslund, J., Cohen, M. J., Jawitz, J. W., et al., 2018. "Solute evidence for hydrological connectivity of geographically isolated wetlands." *Land Degrad. Dev.* 29(11): 3954-3962.

Tiner, R. W., Lang, M. W., Klemas, V. V., 2015. *Remote Sensing of Wetlands: Applications and Advances.* Boca Raton: CRC Press.

UNEP (The United Nations Environment Programme), 2019. Frontiers 2018/19: Emerging Issues of Environmental Concern. Geneva, Switzerland.

UN-Water (The United Nations World Water Assessment Programme), 2018. The United Nations World Water Development Report 2018: Nature-Based Solutions for Water. Paris, UNESCO.

van Griensvan, A., Meixner, T., Grunwald, S., et al., 2006. "A global sensitivity analysis tool for the parameters of multi-variable catchment models." *J. Hydrol.* 324(1-4): 10-23.

Vanderlinden, K., Vereecken, H., Hardelauf, H., et al., 2012. "Temporal stability of soil water contents: A review of data and analyses." *Vadose Zone J.* 11(4):1-19.

Verhoeven, J. T., Arheimer, B., Yin, C., et al., 2006. "Regional and global concerns over wetlands and water quality." *Trends Ecol. Evol.* 21(2): 96-103.

Wang, X., Yang, W., Melesse, A. M., 2008. "Using hydrologic equivalent wetland concept within SWAT to estimate streamflow in watersheds with numerous wetlands." *T. ASAE* 51(1): 55-72.

Whigham, D. F., and Jordan, T. E., 2003. "Isolated wetlands and water quality." *Wetlands.* 23(3): 541-549.

Williams, R., and Fryirs, K., 2016. "Management and conservation of a unique and diverse Australian river type: Chain-of-ponds." The 8th Australian Stream Management Conference, Australia, Jul 31- Aug 3.

Wu, Q., 2018. "2.07-GIS and remote sensing applications in wetland mapping and monitoring." *Comprehensive Geographic Information Systems:*140-157.

Wu, Q., Lane, C. R., Li, X., Zhao, K., et al., 2019. "Integrating LiDAR data and multi-temporal aerial imagery to map wetland inundation dynamics using Google Earth Engine." *Remote Sens. Environ.*228: 1-13.

Wu, Y., Zhang, G., Rousseau, A. N., et al., 2020. "Quantifying streamflow regulation services of wetlands with an emphasis on quickflow and baseflow responses in the Upper Nenjiang River Basin, Northeast China." *J. Hydrol.* 583: 124565.

Xia, Y., She, D., Zhang, W., et al., 2018. "Improving denitrification models by including bacterial and periphytic biofilm in a shallow water-sediment system." *Water Resour. Res.* 54(10): 8146-8159.

Xu, W., Fan, X., Ma, J., et al., 2019. "Hidden loss of wetlands in China." *Curr. Biol.* 29(18): 3065-3071.

Yan, R., Gao, J., Huang, J., 2016. "WALRUS-paddy model for simulating the hydrological processes of lowland polders with paddy fields and pumping stations." *Agr. Water Manage.* 169: 148-161.

Yasarer, L. M., Bingner, R. L., Momm, H. G., 2018. "Characterizing ponds in a watershed simulation and evaluating their influence on streamflow in a Mississippi watershed." *Hydrolog. Sci. J.* 63(2): 302-311.

Yen, H., White, M. J., Arnold, J. G., et al., 2016. "Western Lake Erie Basin: Soft-data-constrained, NHDPlus resolution watershed modeling and exploration of applicable conservation scenarios." *Sci. Total Environ.* 569-570: 1265-1281.

Yeo, I. Y., Lee, S., Lang, M. W., et al., 2019. "Mapping landscape-level hydrological connectivity of headwater wetlands to downstream waters: A catchment modeling approach-Part 2." *Sci. Total Environ.* 653: 1557-1570.

Yin, C., and Shan, B., 2001. "Multi-pond systems: A sustainable way to control diffuse phosphorus pollution." *Ambio* 30(6): 369-375.

Yin, C., Shan, B., Mao, Z., 2006. "Sustainable Water Management by Using Wetlands in Catchments with Intensive Land Use." In *Wetlands and Natural Resource Management. Ecological Studies,* edited by Verhoeven, J.T.A., Beltman, B., Bobbink, R., Whigham, D.F., Berlin: Springer.

Yu, K. J., Jiang, Q. Z., Wang, Z. F., et al., 2015. "The research progress and prospect of Beitang landscape." *Areal Research and Development* 34(3): 130-136.

Yu, L., Rozemeijer, J., Van Breukelen, B.M., et al., 2018. "Groundwater impacts on surface water quality and nutrient loads in lowland polder catchments: Monitoring the greater Amsterdam area." *Hydrol. Earth Syst. Sc.* 22(1): 487-508.

Zhang, H., Huang, G. H., Wang, D., et al., 2012. "An integrated multi-level watershed-reservoir modeling system for examining hydrological and biogeochemical processes in small prairie watersheds." *Water Res.* 46(4): 1207-1224.

Zhang, W., Li, H., Pueppke, S. G., et al., 2020. "Nutrient loss is sensitive to land cover changes and slope gradients of agricultural hillsides: Evidence from four contrasting pond systems in a hilly catchment." *Agri. Water Manage.* 237: 106165.

Appendix

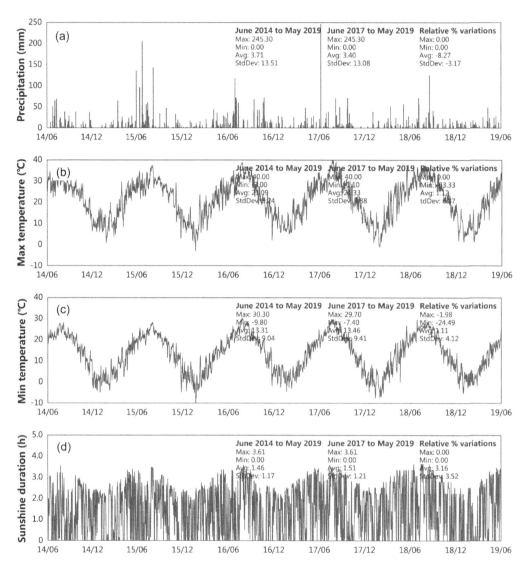

Fig. A4.1 Variations of daily precipitation (a), maximum/minimum temperature (b) and (c), and sunshine duration (d) from June 2014 to May 2019. Each plot contains statistics (maximum, minimum, average value, and standard deviation) and relative percentage changes of the 2 years to past half-decade.

Chapter

5

Attributing Influencing Factors of Water Quality Variation in Multi-pond Systems

5.1 Introduction

Non-floodplain wetlands (NFWs), those surrounded by uplands outside of floodplains and riparian areas, are increasingly recognized for their positive contributions to ecosystems, such as pollutant retention, habitat provision and aesthetics (Golden et al., 2019). These services are largely influenced by water quality within wetlands, and hydrological, biogeochemical and anthropogenic interactions between the wetlands and surrounding landscapes (Lane et al., 2018; Hansen et al., 2018). Understanding water quality variations and prioritizing the influence of extrinsic (outside of the waters) environments and intrinsic (within the waters) characteristics of NFWs is an emerging topic in environmental earth sciences (Usio et al., 2017; Chen et al., 2019). This knowledge also serves as a prerequisite for developing best management practices (BMPs), low impact development, and nature-based solutions, when addressing several technologic and socioeconomic challenges toward sustainable development (Thorslund et al., 2017; UN-Water, 2018).

Multi-pond systems (MPSs) are a common type of NFW in southern China which emerged from traditional water conservancy activities in humid subtropical climates (Verhoeven et al., 2006; Liu et al., 2009). MPSs consist of scattered ponds located among farms and rural households. Ponds form naturally in low-lying parts of the landscape and hold water temporarily. Over time, people dammed and connected such ponds to stabilize water storage. MPSs first appeared 22 centuries ago and currently number approximately 6 million (Yu et al., 2015). These wetland systems gained popularity for multiple reasons including low cost, easy construction, effectiveness for irrigation and drainage, water purification, drinking water, and fishing. Similar NFWs have been reported in other subtropical agricultural landscapes, such as the North Mediterranean Basin (Poschlod and Braun-Reichert, 2017), North American Coastal Plain (Scott et al., 2019), and Southern Highlands of Australia (Williams and Fryirs, 2016), although our focus is on the Chinese context.

Previous studies have demonstrated that MPS water quality is associated with properties of extrinsic environments, i.e., environments that are outside of the ponds themselves, but belonging to their catchments and thus hydrologically connected to the ponds. Examples include land use and land cover properties and impacts of animal feeding (Verhoeven et al., 2006; Xia et al., 2016). Furthermore, impervious rural residential areas can accelerate the transport of nonpoint and mini-point source pollutants to ponded waters (Park et al., 2014), while forests usually act as nutrient buffers due to their fixation and adsorption functions (Lischeid et al., 2018). Meanwhile, pond intrinsic characteristics, such as their water storage capacity, sediment and macrophyte composition, can affect nutrient biogeochemical cycles. For instance, flocculation, precipitation, adsorption and biodegradation can reduce the concentration of particulate organic nitrogen (Li et al., 2019); however, sediment release of ammonium and dissolved phosphorus is significant in deep, anaerobic environments and during heavy rainfall events, resulting in possible role reversals for the MPSs from "sink" from the surrounding farmland to "source" for downstream waters (Nie et al., 2012). Although the extrinsic and intrinsic water quality influences of MPSs have been quantitatively and individually assessed (Xia et al., 2016; Chen et al., 2018; Li et al., 2019; Zhang et al., 2020a), their synergistic (combined) effects at the landscape scale remain largely unquantified. There is thus a need for more studies and quantification methods that can capture the multiple water quality influences and processes among these wetland systems.

Whereas process-based catchment models can have advantageous predictive capabilities, the application of statistical methods, such as multivariate regression, hierarchical clustering, and redundancy analysis, to observational data is fundamental for hypothesis testing with its potential to reveal correlations and overall drivers while using relatively limited computing resources (Astel et al., 2007; Luo et al., 2019; Zhang et al., 2020b). Many of these studies, however, neglect impacts of land-water interactions and fail to distinguish between local aquatic features and features of the surrounding land areas, which are required for management goals (Golden et al., 2017; Hansen et al., 2018). Geographically weighted regression can identify such discrepancies but is mainly applied in large watersheds and river basins and limited by the spatial resolution of input data (Wang and Zhang, 2018). As water quality observations tend to be more stochastic, non-linear and redundant regarding small water bodies, unsupervised clustering is preferred to extract high-dimensional features of these data. Specifically, self-organizing maps (SOM), which belong to the category of competitive learning networks in artificial neural networks, are effective for classifying and visualizing data patterns in a highly variable environment (Vesanto and Alhoniemi, 2000; Céréghino et al., 2009). SOM provides a feasible but rarely tested solution to parse MPS water quality variations for associated source attributions with promising applications in correlating agricultural practices and water chemistry (Romić et al., 2020), selecting biological indicators to reflect environmental degradation (Milošević et al., 2016), and optimizing storm water monitoring programs (Ki et al., 2011).

Over the past several decades, NFWs, such as MPSs, have faced a severe decline in number and in ecosystem service provisioning owing to expanding urbanization, agricultural intensification, and climatic extremes (Marton et al., 2015; Lischeid et al., 2018). While little environmental legislation has focused on pond conservation in China, several policies and regulations have gradually acknowledged their importance, such as restoring clean and healthy catchments in the River Chief Mechanism and pond reinforcement and dredging in Beautiful Rural Development (Dai, 2019). Hence, pond prioritization and management for rural planners and environmentalists also need refined assessments on the water quality variations and influencing factors to achieve the potential benefits of MPSs.

The overall aim of this study is to relate the water quality variations of subtropical agricultural MPSs to their extrinsic and intrinsic influences, including rainfall conditions. Specifically, we want to address the following open scientific questions: (1) how to identify the relative importance of different influences on spatiotemporal variations in water quality? (2) how does rainfall conditions influence the net effects on their water quality improvement or degradation? and (3) to what degree can the water quality variations of these wetland systems be interpreted by the synergistic effects? We address these questions by studying the headwaters of two basins of environmental concern: the Taihu Lake Basin and Qinhuai River Basin (Fig. 5.1b). Six water quality variables in conjunction with surface sediment and macrophyte conditions were examined at these sites. A sequential statistical approach was employed to assess the synergistic effects of intrinsic and extrinsic influences. The SOM as a first step was developed to investigate the spatial and temporal water quality patterns. The linear mixed-effects model (LMEM) as a second step was deployed to relate each of the identified patterns to the intrinsic and extrinsic characteristics of MPSs, followed by discussions on the degree and variability of these influences, as well as implications for rural planning and pond management. The proposed methods and analytical results can be a timely help to identify the biogeochemical hotspots of similar NFWs at local scales and understand pollutant fate and transport in catchments with NFWs. They can also inform management and conservation efforts in the context of global wetland degradation.

5.2 Materials and methods
5.2.1 Study sites and geospatial data

Six MPSs with 3 to 5 ponds each in the Mt. Mao Region (N31°34'-55', E119°10'-23') were selected as study sites. These MPSs were characterized by (1) traditional farming and free-range poultry feeding nearby; (2) upstream location to the reservoirs, where hydrologic regimes are primarily determined by natural rainfall and fill-spill connections; and (3) environmental gradients from low mountains to plains (Fig. 5.1). Both MPS catchment and pond drainage area were delineated from a 5-m resolution DEM (Fig. A5.1a). Pond depth was measured using a boat-mounted ultrasonic device with GPS units, and averaged between 0.4 and 2.1 m. As the average elevation of the 6 MPS catchments dropped from 146 to 32 m, the area proportion of agricultural land (including paddy fields, dry farmland and tea gardens) increased from 17% to 48%, forest (deciduous and evergreen) decreased from 77% to 43%, rural residential areas increased from 1% to 9%, and pond coverage remained at ~ 5%, as extracted from a 0.5-m panchromatic satellite image (Fig. A5.1b). According to local soil surveys (Zhang et al., 1986), Dystric Regosols were predominant in the first 4 catchments but accounted for 40 and 23%, respectively, and shared the remaining area with Eutric Planosols in the last 2 catchments (Fig. A5.1c). Considering the intermittent pollution discharge from mini-point sources (Chen et al., 2018), locations of 20 flocks of chickens and 4 flocks of ducks were estimated (Fig. A5.1d), with surveyed poultry amount in each flock. National Ground Climatological Data (Station No. 58345; N31°26', E119°29', 7.7 m in elevation) showed average annual precipitation from 822-1484 mm. Approximately 70% of this precipitation occurred during the wet season from

Fig. 5.1 Study MPS catchments, including water quality and sediment sampling site locations

April to September and the remainder occurred during the dry season from October to March. Besides, average daily high and low temperatures were 21°C and 13°C over the past 5 years.

5.2.2 Sample collection and analysis

Water quality, sediment and macrophytes were investigated from July 2017 to January 2020, with their corresponding variables, periods and sample sizes shown in Table A5.1. Water samples were collected at a monthly interval near the pond outlet (Fig. 5.2). In the middle of dry and wet seasons, surface sediments (0-15 cm in depth) were seasonally sampled near the above 23 sites and in the middle of the 6 largest ponds for the subsequent average calculation of these waters. Water and sediment samples were both prepared with another duplicate at each site to avoid the risk of randomness. Before lab analysis, water samples were stored in 450 ml polyethene bottles and kept at 4°C, while sediment samples were sealed in foil, stored in 250g PE bags, and then cooled to -20°C. As pond eutrophication can be attributed to external sources, such as fertilization, domestic sewage and animal feeding in agricultural headwaters (Liu et al., 2009; Marton et al., 2015), water quality variables, including NH_4^+-N, NO_3^--N, TN and TP, were selected to represent excessive nutrients, while COD_{Mn} and Chl-a (Chl-a was analyzed bimonthly) were included for organic pollutants and algae, respectively. Similarly, sediment variables contained TN and TP to reflect the general depositional status and total organic carbon (TOC) for the percent fines. The lab analysis on water and sediment samples was conducted within 24 hours and followed SEPA (2002) and Bao (2000), respectively.

A small number of macrophyte species, including lotus (*Nelumbo nucifera*), water caltrop (*Trapa bispinosa*), reed (*Phragmites australis*) and hydrilla (*Hydrilla verticillata*), were found in 17 ponds across the MPSs. Hence, species richness, percentage cover and biomass were selected to reflect the macrophyte conditions of each pond. These variables were measured trimonthly since August 2017, based on field surveys, sample collection and lab analysis. Species richness is the count of all species in the pond. Reckoned via the GPS records and ArcToolbox, percentage cover is the area proportion of macrophyte coverage to the entire pond, which did not distinguish between species. With detailed procedures in Text A5.1, biomass was expressed as dry weight and considered different species and their respective coverage.

5.2.3 Statistical analysis

Using the global Moran's I index, both spatial and temporal autocorrelations were first tested for the differentiation characteristics of above water quality variables. An SOM was then developed to investigate the spatiotemporal patterns, and an LMEM was deployed to link the variations of each identified pattern with pond intrinsic characteristics and extrinsic environments. These methods were further narrated in Text A5.2 with detailed procedures on the latter two parts as follows:

5.2.3.1 Self-organizing map (SOM)

The SOM is an artificial neural network algorithm that can be used to project high-dimensional datasets onto a 2-d, regular map proportional to the degree of similarity (Kohonen, 1982). The main purpose of an SOM is to generate effective and meaningful reference vectors to link input eigenvectors of observations to the output neuron map. The reference vectors can be iteratively optimized through a 3-step training algorithm: competition between the neuron nodes, selection of a winner node, and update of the reference vector of each node. Detailed SOM theories and calculation processes are described in Astel et al. (2007) and Céréghino and Park (2009).

In this study, the 6 water quality variables were used as input eigenvectors after normalization. Three important procedures, including design of the SOM structure (i.e., calculation on the total number of neuron nodes and side lengths of neuron map), selection of proper training methods and data cluster methods, are elaborated. First, the neuron node number was estimated by the heuristic rule $n = 5 \times \sqrt{\text{number of input data}}$ (Vesanto and Alhoniemi, 2000) and a suggested compression range within 30% for values of environmental variables (Kanevski et al., 2009), to balance the accuracy of pattern classification and topographical proximity of the clusters. The map side lengths were determined by the ratio of the two biggest eigenvalues of input data, with some manual adjustments to improve the model performance described below. Second, reference vectors with a hexagonal lattice are set using the linear initialization method. Here, the hexagonal lattice can avoid partiality in the vertical and horizontal directions, while linear initialization can accelerate the next training phase by incorporating prior information of the input data (Astel et al., 2007). Last, partitional clustering and hierarchical clustering were integrated for the fine-tuning of the reference vectors. The former method employs the k-means algorithm, in which the optimal cluster number was selected by the lowest Davies-Bouldin index (Davies and Bouldin, 1979), while the latter method uses the Ward's linkage algorithm to define the cluster boundaries of neuron nodes (Nguyen et al., 2015).

QE (quantization error) and TE (topology error) were used to evaluate the model performance. QE measures map resolution by averaging the distance between each data vector and its best matching unit (BMU). TE provides a measure of map accuracy by calculating the proportion of all data vectors for which 1st and 2nd BMUs are not adjacent. Further descriptions and calculation formulas of the QE and TE can be found in Ki et al. (2011) and Park et al. (2014). Note that both of the criteria gradually decrease with increasing map size (Céréghino and Park, 2009), so the optimization of the number of neuron nodes and side lengths of neuron map requires manual adjustments to locate the minimum values of criteria under the heuristic rule and suggested compression range.

5.2.3.2 Linear mixed-effects model (LMEM)

The LMEM was used to identify and quantify the relationship between MPS water quality and potential influencing factors in each spatiotemporal pattern, after testing the normal distribution of the response variables ($p = 0.05$) using the Ryan-Joiner Normality test (Nosakhare and Bright, 2017), and log-transforming those required to meet normal distribution assumptions. The

LMEM, proposed by Laird and Ware (1982), can provide a robust, simultaneous evaluation of the associations between response variables and environmental gradients, while considering the repeated observations in input data (Lessels and Bishop, 2013; Xia et al., 2016). Detailed LMEM theories and comparisons with traditional regressive techniques are in Pinheiro and Bates (2000) and Bolker et al. (2009).

In this study, pond intrinsic characteristics, including average depth, sediment and macrophyte status, and extrinsic catchment environments, including average slope, land uses, soil properties and poultry feeding were used as candidate independent variables in the LMEM. The land use variables consisted of area proportion of water (i.e., pond), forest, agricultural land, and rural residential area. The soil properties were reflected by the area proportion of Dystric Regosols due to limited soil types. Poultry feeding was reckoned by the total number of chickens and ducks in all flocks, as the nutrient export of both animals is on the same order of magnitude and is usually aggregated in coefficient estimation (Li et al., 2016). The sediment and macrophyte status corresponded to their respective variables in Table A5.1. Pond water storage and fill-spill connections were not explicitly included, as the former can be reflected by the average depth and area proportion, while the latter, including serial or parallel in structure and perennial or intermittent in continuity, was generally associated with the land uses between up- and downstream ponds (e.g., a more interlaced ditch network in agricultural land according to Liu et al., 2009 and Chen et al., 2020).

As an intermediary between extrinsic and intrinsic characteristics, rainfall was expected to influence the fate and transport of nutrients from terrestrial to aquatic environments (Ouyang et al., 2015; Lane et al., 2018). Hence, the statistical interaction of monthly accumulative rainfall (considering our sampling intervals) and each of the above factors was incorporated via their cross product, allowing the effect of that factor to change linearly with rainfall (i.e., enhanced or attenuated by rainfall). Thus, for each spatiotemporal pattern of water quality variations, the water quality variables, potential influencing factors, and rainfall effects were preliminarily linked below (see Text A5.3 for interpretation of the symbols), after normalizing the independent variables:

$$
\begin{aligned}
\text{Concentration} = {} & \text{Intercept} + a{\cdot}\text{Slope} + b{\cdot}\text{Landuse} + c{\cdot}\text{Soil} + d{\cdot}\text{Poultry} + \\
& e{\cdot}\text{Depth} + f{\cdot}\text{Sediment} + g{\cdot}\text{Macrophyte} + h{\cdot}\text{rainfall} + A{\cdot}(\text{Slope} \times \text{rainfall}) + \\
& B{\cdot}(\text{Landuse} \times \text{rainfall}) + C{\cdot}(\text{Soil} \times \text{rainfall}) + D{\cdot}(\text{Poultry} \times \text{rainfall}) + \\
& E{\cdot}(\text{Depth} \times \text{rainfall}) + F{\cdot}(\text{Sediment} \times \text{rainfall}) + \\
& G{\cdot}(\text{Macrophyte} \times \text{rainfall}) + \varepsilon
\end{aligned}
\tag{1}
$$

A statistical power analysis was performed to reduce model complexity and ensure variable interpretability. With detailed descriptions in Text A5.4, this method was reversely used to determine the maximum acceptable number of independent variables, after identifying the water quality samples of each LMEM. Thus, candidate variables with minimum standard deviation were iteratively abandoned with their rainfall interactions from Eq. (1), as these variables are more likely to be covered by Intercept and tend to be less sensitive than other ones following Shen et al., 2014 and Zhang et al., 2020a. The coefficient of determination (R^2) and simple linear regression against observations and simulations were constructed to evaluate the goodness-of-fit of the LMEM. According to previous NPSP simulations (Yen et al., 2016; Chen et al., 2018; Li et al., 2019), the R^2 values were considered "very good", "good", and "fair", when reaching 0.8, 0.7, and 0.6, respectively. The statistically significant variables and their coefficients were then screened out to assess the direction and degree of each influencing factor.

5.3 Results
5.3.1 Spatial and seasonal patterns of water quality

Table A5.2 shows the result ranges of autocorrelation analysis. Comparing with previous SOM and LMEM applications (Jin et al., 2011; Lessels and Bishop, 2013; Park et al., 2014; Romić et al., 2020), our water quality samples presented weak to moderate spatial autocorrelation but slightly higher temporal autocorrelation. Hence, the spatiotemporal proximity does not impede the identification of water quality patterns, although our study scale was much smaller than that of large watersheds and river basins.

Using the monthly observations at 23 sites from July 2017 to January 2020 as input eigenvectors, the QE and TE, with final values of 0.870 and 0.019, respectively, indicated a well-trained SOM in topology (Céréghino and Park, 2009). Following the output SOM (Fig. A5.2) and unsupervised clustering of water quality observations (Fig. A5.2), Fig. 5.2 presents the average concentration and standard deviation of the water quality variables in each of the 6 clusters. The significant differences in these clusters again demonstrated the effectiveness of the SOM analysis. To be specific, clusters No. 1 and 2 had more variables with high average values, where No. 1 peaked for NO_3^--N, TN, and TP, and No. 2 peaked for NH_4^+-N and Chl-a. Clusters No. 3 and 5 had fewer variables with high and medium average values, but more variables with low average values. The minimum average concentration of NH_4^+-N, COD_{Mn} and Chl-a, for example, were in cluster No. 5. Clusters No. 4 and 6 hold a similar number of variables with low average values, where the former was minimized in TP, and the latter minimized in TN.

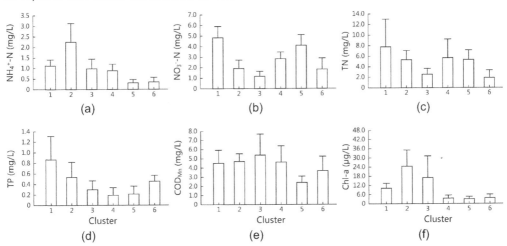

Fig. 5.2 Average concentration and standard deviation of the water quality variables in the SOM clusters

To parse the above discrepancies in space and time, water quality observations of the unsupervised clusters were traced back to their original sampling sites and times. Using the Jenks natural breaks of DEM and relative pond positions in the catchment (i.e., up-, mid- and downstream) for spatial grouping, and seasons (i.e., wet and dry) for temporal grouping, water quality patterns of the 6 clusters were exhibited in Fig. 5.4. Clusters No. 1, 4, and 5 were mostly sampled in the upstream ponds. While the samples of cluster No. 1 were almost equal in the wet and dry season, No. 4 had more in the wet season, and No. 5 had a majority in the dry season. Clusters No. 2 and 3 held the largest samples in the midstream ponds, in which ~ 70% were collected in the wet season. Cluster No. 6 was almost halved in the mid- and downstream ponds, in which samples of the dry season were a bit more than those of the wet season.

Fig. 5.3 Spatial and seasonal patterns of the MPS water quality in the unsupervised SOM clusters. Plots (a) to (f) correspond to clusters No. 1 to 6, respectively. The spatial and seasonal occurrence represents the total number of water quality samples in each pond and season, respectively.

5.3.2 Characteristics of extrinsic and intrinsic factors

Table 5.1 presents characteristics of the extrinsic environment of the up-, mid-, and downstream ponds. As the average slope of the pond drainage area decreased from 7.5 to 5.3°, the average area proportion of forest dropped from ~ 69 to 56%, and agricultural land increased from ~ 24 to 31%. The water coverage was almost flat at ~ 5% in the up- and midstream, but increased to ~ 8% in the downstream, while rural residential areas increased from ~ 3% in the upstream to ~ 5% in the mid- and downstream. Fully occupying the upstream, Dystric Regosols had a decreasing average area proportion but increasing standard deviation in the mid- and downstream. Moreover, poultry feeding existed across the elevation gradient, where the average total amount was markedly higher in the midstream than that in the up- and downstream. The identified discrepancies revealed a system gradually shifting from natural forest to agriculture and rural life, which reflects MPSs' extrinsic environments in the transition zone from low mountains to plains (Chen et al., 2019; Li et al., 2019).

Similarly, Table 5.2 compares the pond intrinsic characteristics of the spatial groups. The average depth dropped from 1.8 m in the upstream to ~ 1.1 m in the mid- and downstream. Such variation, in conjunction with the average slope and pond area proportion of the drainage area, implied that the upstream ponds were more functional in floodwater storage, while mid- and downstream ones were closer to irrigation and rural life, as summarized by Yu et al. (2015). Meanwhile, significant differences were found in surface sediments and macrophytes. In particular, the average concentration of TN was higher in the up- and midstream than that in the downstream (~ 2000 vs. 1,600 mg/kg). Conversely, the average TP level was higher in the upstream than that in the mid- and downstream (335 vs. ~ 210 mg/kg), while TOC content was almost unchanged at ~ 1.3%. When the average number of macrophyte species increased slightly (0.8 to 1.3) and biomass increased markedly downgradient (~ 60 to 140 g/m2), the percentage cover was generally higher in the midstream (14%) than in other areas (~ 11%). Although pond sediment showed spatial variability, macrophytes were assumed to be more natural in the upstream and prone to human planting and maintenance in the mid- and downstream.

Table 5.1 Variations of the extrinsic environment across the up-, mid-, and downstream ponds

Category	Factor (unit)	Upstream		Midstream		Downstream		Adopted [b]
		Range	Mean (Stdev [a])	Range	Mean (Stdev)	Range	Mean (Stdev)	
Physical property	Average slope (°)	1.6 – 12.3	7.5 (4.3)	1.5 – 11.0	5.7 (4.0)	1.4 – 11.6	5.3 (3.4)	
Land use	Water (%)	1.7 – 11.7	4.6 (3.7)	2.8 – 9.5	4.8 (2.5)	1.2 – 23.3	8.6 (8.8)	√
	Forest (%)	43.5 – 89.8	69.1 (17.9)	39.6 – 76.2	60.7 (15.4)	0 – 89.7	56.3 (29.6)	√
	Agricultural land (%)	5.8 – 47.1	23.8 (15.9)	11.9 – 52.2	27.5 (16.6)	9.1 – 66.4	30.8 (19.7)	√

Continued

Category	Factor (unit)	Upstream Range	Upstream Mean (Stdev [a])	Midstream Range	Midstream Mean (Stdev)	Downstream Range	Downstream Mean (Stdev)	Adopted [b]
Land use	Rural residential areas (%)	0 – 6.5	3.2 (2.9)	2.3 – 9.5	5.1 (2.8)	0 – 12.8	5.4 (4.7)	√
Soil	Dystric Regosols (%)	100 – 100	100 (0.0)	7.8– 100	78.1 (30.5)	0 – 100	66.5 (44.6)	√ [c]
Poultry feeding	Chickens and Ducks (#)	0 – 92	29 (16.2)	0 – 167	38 (35.3)	0 – 122	25 (22.2)	√

[a] Standard deviation.
[b] This column shows the adopted LMEM independent variables for the Section 5.3.3.
[c] This variable and its rainfall interactions were only adopted for the mid- and downstream LMEMs.

Table 5.2 Variations of the intrinsic characteristics across the ponds' spatial groups. Footnotes are the same to Table 5.1

Category	Factor (unit)	Upstream Range	Upstream Mean (Stdev [a])	Midstream Range	Midstream Mean (Stdev)	Downstream Range	Downstream Mean (Stdev)	Adopted [b]
Physical property	Average depth (m)	0.7 – 2.3	1.8 (0.7)	0.7 – 1.8	1.2 (0.5)	0.4 – 1.6	1.1 (0.4)	
Sediment	TN (mg/kg)	1484 – 2435	1932 (338)	1,721 – 2,562	2,086 (258)	1,284 – 2,004	1,615 (186)	√
Sediment	TP (mg/kg)	241 – 397	335 (47)	123 – 298	198 (56)	165 – 289	231 (38)	√
Sediment	TOC (%)	1.14 – 1.44	1.30 (0.10)	1.15 – 1.54	1.28 (0.11)	1.17 – 1.46	1.27 (0.07)	
Macrophytes	Species richness (#)	0 – 1	0.8 (0.4)	0 – 3	1.0 (1.1)	0 – 3	1.3 (1.0)	
Macrophytes	Percentage cover (%)	0 – 28.4	10.1 (7.8)	0 – 42.5	13.7 (14.7)	0 – 53.7	12.3 (13.6)	√ [b]
Macrophytes	Biomass (g/m2)	0 – 197.1	59.2 (57.2)	0 – 487.6	128.4 (142.3)	0 – 460.3	141.2 (134.5)	√

Seasonal changes of sediment and macrophytes were highlighted for the 23 ponds. The time-variability of extrinsic influences was not included, because the proportion of land use types remains almost unchanged across the study period, while increased nutrient export due to poultry growth in a year is roughly offset by local consumptions (Thornhill et al., 2018). According to Fig. 5.4, both max and average concentrations of TN and TP in surface sediments decreased slightly from dry to wet seasons, presumably due to the higher suspension and lower sedimentation effects during rainfall-runoff processes. The percentage of TOC, however, levelled off across the seasons. For pond macrophytes, max and average levels of percentage cover and biomass were markedly higher in the wet season than those in the dry season, despite a slight decrease for the biomass at the end of the wet season. These variations coincided with the general knowledge that the simultaneity of heat and water can strongly accelerate the macrophyte growth in shallow waters during the wet season in subtropical climates (Kosten et al., 2009).

Fig. 5.4 Variations of sediment and macrophytes for the 23 ponds across the 2.5-year experimental period

5.3.3 Correlation between water quality and influencing factors

The last column of Table 5.1 and 5.2 shows the adopted LMEM independent variables, after identifying water quality samples for the up-, mid-, and downstream simulations (n = 90, 120, 135, respectively), and performing the corresponding power analysis. Hence, there were finally 17, 21, and 21 coefficients for the 3 spatial groups, respectively (see Text A5.3 for the initial numbers). Grouped LMEM simulations for ponds in 3 relative positions and 2 seasons were generally satisfactory (Fig. A5.4 to A5.6), as 66.7% of the R^2 values were "very good" and "good", and 27.8% reached and 5.5% was slightly lower than the "fair" level. Besides, the model residuals were all normally distributed, with ranges slightly enlarged as the water quality observations increased (see Text A5.5 for further comparisons and interpretations). Note that 11 out of the 36 water quality response variables were log transformed (Table A5.3), according to their normal distribution tests. The adopted independent variables were normalized before analysis.

Table 5.3 presents the significant independent variables ($p < 0.05$) and their coefficients in the wet and dry seasons, which gradually improved LMEM performance downgradient. For the upstream ponds, "%Agricultural land", "%Rural residential areas", and their rainfall interactions exhibited consistent, drastic positive effects on the water quality variables. "%Forest", "%Water", and their rainfall interactions, however, presented both positive and negative effects. For example, "%Forest" can decrease NH_4^+-N in the wet season, but "%Forest × Rainfall" can increase NO_3^--N in the wet season and Chl-a in both seasons. Similarly, "Rainfall" had a slight positive effect on NH_4^+-N in the dry season, but its negative effect was stronger for COD_{Mn} in both seasons than that for TP in the wet season. Meanwhile, "#Poultry" and its rainfall interactions can increase TP and COD_{Mn} levels, especially in the wet season. While nitrogen concentrations of pond sediment mostly had intrinsic, positive effects during rainfall-runoff processes, e.g., "%TN × Rainfall" for NO_3^--N and TN in both seasons and for COD_{Mn} in the dry season, the macrophyte status, i.e., "Biomass", can reduce NH_4^+-N and Chl-a in the wet season.

Table 5.3 Significant independent variables (p < 0.05) and their coefficients of the LMEM analysis (see Text A5.6 for interpretation of the abbreviations)

Response variable	Upstream				Midstream				Downstream			
	Wet season		Dry season		Wet season		Dry season		Wet season		Dry season	
	Coefficient	Effect	Coefficient	Effect	Coefficient	Effect	Coefficient	Effect	Coefficient	Effect	Coefficient	Effect
NH_4^+-N[a]	2.51	%Resi	1.61	%Resi	–	–	–	–	–	–	–	–
	1.36	%Agri	0.41	Rain	–	–	–	–	–	–	–	–
	-0.21	Biomass	0.17	#Poultry	–	–	–	–	–	–	–	–
	-0.84	%Forest	-1.43	%Water	–	–	–	–	–	–	–	–
	-1.77	%Water × Rain	–	–	–	–	–	–	–	–	–	–
NO_3^--N	7.24	%Agri	4.65	%Agri	2.05	%Agri × Rain	0.47	%Agri	1.98	%Resi × Rain	1.03	%Resi
	2.60	%Resi × Rain	1.94	%Resi	0.38	%Resi × Rain	0.39	TN × Rain	0.42	%Agri × Rain	0.40	TN
	0.46	TN × Rain	0.36	TN × Rain	0.10	%Forest	-0.79	Biomass	0.21	TN × Rain	0.11	%Water
	0.11	%Forest × Rain	–	–	-1.81	%Water × Rain	–	–	-0.72	Rain	0.07	%Cover
	-2.28	%Water	0.18	%Water × Rain	–	–	–	–	–	–	–	–
TN	4.65	%Agri	1.22	%Agri	2.67	%Agri × Rain	1.31	%Agri × Rain	–	–	–	–
	0.77	%Resi × Rain	0.94	%Resi	0.29	%Resi	0.62	%Resi × Rain	–	–	–	–
	0.53	TN × Rain	0.23	TN × Rain	0.04	#Poultry	0.16	TN × Rain	–	–	–	–
	–	–	0.18	%Water × Rain	-0.16	%Cover	–	–	–	–	–	–
TP[a]	0.62	%Agri × Rain	0.53	%Agri	0.27	%Agri × Rain	0.46	%Agri	–	–	–	–

Response variable	Upstream				Midstream				Downstream			
	Wet season		Dry season		Wet season		Dry season		Wet season		Dry season	
	Coefficient	Effect	Coefficient	Effect	Coefficient	Effect	Coefficient	Effect	Coefficient	Effect	Coefficient	Effect
TP[a]	0.27	#Poultry × Rain	0.11	%Resi	0.14	#Poultry × Rain	0.11	%Resi	-	-	-	-
	0.16	%Water × Rain	-0.19	%Water	0.10	%Rego × Rain	0.04	TP × Rain	-	-	-	-
	0.07	%Resi	-	-	-0.56	%Water × Rain	-0.53	%Forest	-	-	-	-
	-0.38	Rain	-	-	-	Biomass	-	-	-	-	-	-
CODMn[a]	2.46	%Resi	1.07	%Resi	1.48	%Resi	1.46	%Agri	3.25	%Resi × Rain	1.18	%Agri
	0.98	%Agri × Rain	0.42	%Agri	0.34	%Agri	0.57	%Resi	1.11	%Agri × Rain	0.31	%Resi
	0.57	#Poultry × Rain	0.13	TN × Rain	0.21	#Poultry	0.33	#Poultry × Rain	0.32	#Poultry × Rain	0.08	TN
	-1.72	Rain	-1.51	Rain	-1.19	Rain	-1.47	Rain	-0.78	Biomass	-0.13	Biomass
	-	-	-	-	-2.76	Biomass	-	-	-	-	-0.46	Rain
Chl-a[a]	0.54	%Agri	0.37	%Agri	0.61	%Agri	0.22	%Agri	0.72	%Resi	0.27	%Resi × Rain
	0.18	%Resi	0.14	%Resi × Rain	0.14	%Resi	0.08	%Resi	0.13	%Agri	0.05	%Agri × Rain
	0.13	%Forest × Rain	0.08	%Forest × Rain	0.08	TN	0.06	Forest × Rain	0.06	TN × Rain	-0.09	%Cover
	-0.41	%Water × Rain	-	-	-0.70	%Water × Rain	-	-	-0.19	%Water × Rain	-	-
	-0.69	Biomass	-	-	-1.74	Biomass	-	-	-0.48	Biomass	-	-

[a] Note that the following 7 response variables were log transformed: NH_4^+-N in the upstream for the wet season, TP in the upstream for both seasons and in the midstream for the dry season, COD_{Mn} in the mid- and downstream for the dry season, and Chl-a in the midstream for the wet season.

For the midstream ponds, "%Agricultural land", "%Rural residential areas", and their rainfall interactions were listed as top contributors to water quality degradation. "%Water × Rainfall" can substantially dilute NO_3^--N, TP, and Chl-a concentrations in the wet season, while "%Forest" and "%Forest × Rainfall" can significantly buffer TP in the dry season, despite their slight source effects for NO_3^--N and Chl-a. "#Poultry" and its rainfall interactions were weakly related to COD_{Mn} increase in both seasons and TN and TP increase in the wet season, while "Rainfall" alone had a strong buffer effect for COD_{Mn}, and "%Dystric Regosols × Rainfall" only and slightly increased TP levels here in the wet season. Like the upstream simulations, "TN" and "TP" of surface sediments acted as a source of NO_3^--N, TN, TP, and Chl-a, especially during rainfall in the dry season. Furthermore, "Biomass" and "%Cover" dilute NO_3^--N in the dry season and TN, COD_{Mn}, and Chl-a in the wet season.

For the downstream ponds, the positive effects of "%Agricultural land" and "%Rural residential areas" remained significant for the NO_3^--N, COD_{Mn}, and Chl-a simulations, which surpassed their adverse effects for the midstream waters. Although "%Water", "Rainfall", and their combinations mostly contribute to the dilution of NO_3^--N, COD_{Mn}, and Chl-a, "%Water" alone slightly increased NO_3^--N in the dry season, and "#Poultry × Rainfall" slightly increased COD_{Mn} in the wet season. Furthermore, "TN" and "TN × Rainfall" were potential sources of the identified water quality variables, while the buffer functions of "Biomass" for COD_{Mn} and Chl-a in the wet season were stronger than that of "%Cover" for NO_3^--N and Chl-a in the dry season.

5.4 Discussion
5.4.1 Extrinsic environments vs. intrinsic characteristics

NFWs, such as prairie potholes, vernal pools and cypress domes, provide numerous ecosystem services via healthy aquatic environments (Golden et al., 2017; Scott et al., 2019). Among these wetlands systems, MPSs are often dominant on small farms in southern China but are rarely assessed in terms of both extrinsic and intrinsic influences on water quality dynamics (Liu et al., 2009; Yu et al., 2015; Xia et al., 2016; Li et al., 2019). In this context, a combination of SOM and LMEM revealed that the synergistic effects of both within-wetland and surrounding landscape processes can account for 69% ± 12% and 73% ± 10% of the variations in the wet and dry season, respectively, following the degree of explanation (i.e., R^2 values) in Fig. A5.4 to A5.6. When averaging the coefficients of Table 5.3 on the weight of their sample size, surrounding external landscape characteristics, in absolute terms, had 1.4, 6.9, 3.2, and 4.3 times more impacts than internal wetland specific characteristics, on elevated levels of NH_4^+-N, NO_3^--N, TP, and COD_{Mn}, respectively. On the other hand, for Chl-a, both wetland internal characteristics and the surrounding landscape contributed equally to observed levels. Note that "Rainfall" was not included here due to the intermediary role, and "%Water" was counted as an intrinsic factor due to its closer relationship with aquatic environments. The relevance of considering the whole wetlandscape, i.e. the larger coupled system of multiple wetlands and their interaction with, and influence by, the surrounding landscape, has previously been highlighted (Thorslund et al., 2017). Our results presented the importance of the larger wetlandscape system, over individual wetland characteristics, in regulating several critical water quality indicators, which can be a timely help to identify and understand pollutant fate and transport in MPSs and similar NFWs.

Previous studies have identified fertilization and domestic sewage as the major sources of aquatic nitrogen and phosphorus pollution (Thornhill et al., 2018; Boardman et al., 2019). This is consistent with our findings in MPSs, as quantified by the coefficient of "%Agricultural land" and "%Rural residential areas" (Table 5.3). In particular, the higher concentrations of NO_3^--N, TN and TP in clusters No. 1 and 5 were largely due to intensive tea plantations in the up- and midstream (Fig A5.1b), where local farmers tend to apply cheap but excessive fertilizer to increase yields (Yan et al., 2018). Meanwhile, agronomic practices, such as trimming old stems and leaves to force new growth, can lead to an abrupt reduction of vegetation coverage at 50% ± 20% and increased vulnerability to soil erosion (Zhang et al., 2020a). Organic pollutants and algae, reflected by COD_{Mn} and Chl-a, were abundant in clusters No. 2 and 3 in the midstream, where rural households were as common as the downstream (Table 5.1), but domestic sewage treatment facilities were mostly absent. These facilities, designed to meet the National Standard of Wastewater Discharge (GB18918-2002, i.e., COD_{Mn}, TN and TP emissions≤60, 20, and 3 mg/L, respectively), are promoted widely in large rural settlements, but are poorly implemented in areas of scattered households at higher elevations due to cost and maintenance issues (Chen et al., 2017; Dai et al., 2019). Moreover, free-range poultry feeding, a smaller but non-negligible source of NH_4^+-N, TP, and COD_{Mn} (Table 5.3), is more developed in the up- and midstream clusters, e.g., duck flocks in ponds No. 5, 10, and 20, and chicken flocks near ponds No. 13, 16, and 21 (Fig A5.1d). This mini-point, intermittent pollution discharge partly explained the in-between trend of NH_4^+-N in the SOM analysis (Fig. A5.2), and partly the higher NH_4^+-N, TP, and COD_{Mn} levels of up- and midstream ponds in clusters No. 1 to 4 (Fig 5.3 and 5.4). Different from the positive correlation between soil erodibility and water quality degradation in the red soil hilly regions of southern China and Corn Belt in the midwestern United States (Borrelli et al., 2017), the variability of soil textures, represented by "%Dystric Regosols", had little influence on pond nutrient dynamics, presumably due to the limited soil types in our small-scale assessment, and the sediment retention capacity of tributary ditches among the fields (Dollinger et al., 2015).

Internal nutrient load from sediment is closely associated with eutrophication of large, shallow aquatic systems, such as Lakes Taihu, Arresø and Clear (Qin et al., 2020). Although surface sediments exhibited consistent, positive effects across the water quality variables, internal nitrogen and phosphorus sources with coefficients ranging from 0.04 to 0.53 were far weaker than the external ones across agricultural MPSs (Table 5.3). This is presumably due to pond dredging that is easy to implement and can reduce blockages and increase farmland fertility (Mitsuo et al., 2014; Fu et al., 2018). Ten among the 13 significant, sediment-related factors were enhanced by the monthly cumulative rainfall, which confirmed the perturbation and re-suspension effects of rainfall-runoff processes (Thornhill et al., 2018; Jin et al., 2020). On the contrary, biological uptake is a crucial approach to improve pond water quality. The aggregated variable, "Biomass", was effective to lower NH_4^+-N, NO_3^--N, COD_{Mn} and Chl-a levels, with functions more pronounced in the mid- and downstream during the wet season. While such spatially differentiated water quality beneficial effects are consistent with the overall macrophyte status in Table 5.2, the seasonal variations can be largely attributed to the accelerated growth in hydrothermally sufficient conditions of humid subtropical climates (Fig. 5.5). Moreover, the intrinsic characteristics can affect nutrient biogeochemical cycles, e.g., ammonification and subsequent nitrification were more microbially active in shallow and aerobic conditions, such as hyporheic and riparian zones (Haygarth et al., 2014). Indicated by the variable "%Cover", which increased as the average depth drops, these processes were confirmed by the degradation of TN and release of NO_3^--N in mid- and downstream ponds.

5.4.2 Spatial and seasonal variations of influencing factors

Generally, land uses with intensive anthropogenic activities are recognized as "sources" of pollutants, while natural landscapes, e.g., waters and forests, are considered pollution "sinks" (Chen et al., 2008; Park et al., 2014; Shen et al., 2014). The "source-sink" framework is beneficial to grasp the key processes in pollutant fate and transport but remains controversial regarding different study characteristics, spatiotemporal scales, and water quality variables (Tu et al., 2013; Xia et al., 2016; Romić et al., 2020). In this study, the influencing direction and degree of some extrinsic factors varied in response to relative pond positions and seasons (Table 5.3). For example, "%Water" is generally related to water quality improvement. Combined with rainfall, such influences on TN and TP, however, were obviously reversed in the headwater (i.e., upstream), where the average slope was higher (Table 5.2) and tributary ditches were less (can be inferred from higher "%Forest" and lower "%Agricultural land" according to Liu et al., 2009 and Chen et al., 2020). Hence, the spatially differentiated effects of "%Water" were largely due to successive, intensified rainfall-runoff-infiltration processes at higher elevations, where nutrient export (especially TN and TP) had reached the buffering limits of downgradient ponds (Lischeid et al., 2018; Chen et al., 2019), resulting in a "sink" to "source" reversal of these waters.

The area proportion of forest had negative influences on NH_4^+-N levels in the upstream and TP levels in the midstream. The NO_3^--N and Chl-a dynamics, however, presented positive correlations with "%Forest" and its rainfall interactions in the same region. These phenomena were presumably associated with increased NO_3^--N leaching due to (1) the pervasive transplant of mature trees for economic purposes (Dirnböck et al., 2016), (2) the less nutrient absorption efficiency of some old trees (Yanai et al., 1995), and (3) more atmospheric nitrogen deposition from tourism and automobile exhaust in low mountains and hills (Qiao et al., 2018); as well as the significant positive correlation between NO_3^--N and Chl-a in aquatic systems, due to the NO_3^--N uptake of phytoplankton in the synthesis of amino acids for their cells (Dortch et al., 1990).

Rainfall is widely accepted as the main trigger of NPSP in subtropical climates. Previous studies usually aimed at the surface and subsurface flow at the hillslope scale (Yan et al., 2018), or streamflow at the catchment scale, to explore the kinetic mechanisms of rainfall-runoff on water quality variations. We linearly attached rainfall effects to the extrinsic and intrinsic characteristics of MPSs. Consistent with previous findings (Lane et al., 2018; Jin et al., 2020), rainfall presented adverse effects when associated with impervious land uses (e.g., "%Rural residential areas"), dilution effects when combined with "%Water" (except upstream TN and TP), and overall washing effects on the parameter itself. Such washing effects, however, varied in degree across relative pond positions and seasons (Table 5.3). In addition to internal nutrient biogeochemical cycles (e.g., nitrification and denitrification), possible explanations include disparate peripheral landscape characteristics (Wang and Zhang, 2018; Boardman et al., 2019), and increasingly complex and interlaced ditch networks downgradient (Chen et al., 2020). Covered up by the area proportion of land use types, these deeper, non-linear rainfall effects were currently inexplicit. Refined modelling on landscape patterns (e.g., compositions, fragmentation and diversity), agronomic practices (e.g., fertilization, crop rotation, irrigation and harvest), and water quality and streamflow dynamics of ditch networks can address these issues to advance MPS research, as indicated by the highest average coefficient of "%Agricultural land".

5.4.3 Implications for rural planning and pond management

Many nations have set a no-net-loss target in wetland management, but basically ignore NFWs, which resulted in hidden loss of wetland functions behind the reported "paper offsets" (Xu et al., 2019). Our SOM analysis highlights the MPS water quality status: (1) TN always exceeded the threshold of Level V Water (2.0 mg/L specified by MEP, 2002), while NH_4^+-N and TP reflected similar degradation ($\geq$ 2.0 and 0.4 mg/L, respectively) in the up- and midstream ponds throughout the year and downstream ponds during the dry season; and (2) Chl-a variations exhibited a hypertrophic state ($\geq$ 28 μg/L) in the midstream ponds during the wet season, according to the Trophic State Index and grading standard in Carlson et al., 1977. Comparing with relevant pond studies in England (soluble reactive phosphorus averaged at 0.38 mg/L, Williams et al., 2010), Denmark (TN and TP averaged at 1.9 and 0.09 mg/L, respectively, Sønderup et al., 2016), and Japan (TN and TP averaged at 1.32 and 0.13 mg/L, respectively, Usio et al., 2017), our findings reflect a more serious situation under fragmented agricultural landscapes and humid subtropical climates of southeast China (Fig. 5.1). They also demonstrate a general and pressing issue, as these small water bodies usually locate in critical source areas of downstream lakes and rivers (Cohen et al., 2016; Li et al., 2019), but the water quality is degraded due to expanding urbanization and agricultural intensification (Chen et al., 2019; Golden et al., 2019). Identifying the variables that reflect contamination status (NH_4^+-N, TN, TP, and Chl-a) and associated spatial and seasonal hotpots (in the mid- and upstream during the wet season) can provide critical information for refined watershed management and pollution reduction.

BMPs in catchments with MPSs should incorporate spatially explicit measures following the identified influences and coefficients (Table 5.3). Measures like precision fertilization and recycling of crop residues are necessary to improve fertilizer use efficiency across the entire catchment. Design of domestic sewage treatment must either cover the upstream areas, or include distributed, portable facilities, such as waterless toilets and mini-septic tanks, as has been tested in farmer-participatory rural environmental reform in Chen et al., 2017. As poultry feeding has been organized into concentrated, fenced areas (Zou et al., 2020), pilot programs and demonstration sites should be prioritized in the up- and midstream. Meanwhile, focus should be placed on MPS intrinsic factors. Pond dredging needs encouragement and revitalization during agricultural slack seasons, as they are cost-efficient, easily implemented to transfer sediment pollutants into farmland fertility, but increasingly ignored due to the limited short-term economic benefit (Mitsuo et al., 2014; Yu et al., 2015). Also, pond area proportion, macrophyte biomass and routine maintenance (plant and harvest), must be guaranteed, especially at higher elevations, as these indicators are effective in nutrient trapping, but vulnerable to human interference (filling, occupying, and animal feeding) and climatic extremes (flood and drought).

Following the novel insights on the original "source-sink" framework, the spatial distribution of ponds, forest, and their collocation with surrounding landscapes must be considered when elaborating sustainable rural planning and pond conservation programs. Process-based models, including watershed models (e.g., SWAT and HSPF), hydrodynamic and water quality models (e.g., WASP and EFDC), and groundwater models (e.g., MODFLOW and

HydroGeoSphere) (see Fu et al., 2019 for detailed model descriptions), can be useful, especially when high-resolution (e.g., using unmanned aerial vehicle and microsatellite platforms) and diverse monitoring data (e.g., about intermittent pond-ditch connections) are available (Lane et al., 2018; Golden et al., 2019). As well-focused MPS modelling involves both surface and subsurface hydrologic processes, and biogeochemical cycles on lands, in ponds and ditches, appropriate coupling of these models, considering the trade-offs between complexity and fidelity (i.e., controllable uncertainties in future changes of land use and pond management), is expected to deep explore the time window and nutrient flux of identified "source-sink" dynamics.

5.5 Conclusions

The small size, wide distribution, and degraded water quality make MPSs a front-line but understudied topic of NPSP. Using a combination of SOM and LMEM, this study serves as a first exploration to link MPSs' water quality variation with their extrinsic and intrinsic influences. Across the 6 test MPSs (comprising 23 ponds), NH_4^+-N, TN, and TP almost always exceeded the surface water quality standard (2.0, 2.0, and 0.4 mg/L, respectively) in the up- and midstream ponds, while Chl-a exhibited hypertrophic state ($\geq$ 28 µg/L) in the midstream ponds during the wet season. The synergistic influences explained 69%±12% and 73%±10% of the water quality variations in the wet and dry season, respectively. The adverse, extrinsic influences were generally 1.4, 6.9, 3.2, and 4.3 times of the beneficial, intrinsic influences for NH_4^+-N, NO_3^--N, TP, and COD_{Mn}, respectively, although the effects of forest and water area proportion presented spatiotemporal variability. While COD_{Mn} was more linked with rural residential areas in the midstream, higher concentrations of TN and TP in the up- and midstream can be attributed to agricultural land, and NH_4^+-N variations had a smaller but non-negligible source of poultry feeding. Pond sediment exhibited consistent adverse effects with amplifications during rainfall-runoff processes, while macrophyte biomass reflected biological uptake of COD_{Mn} and Chl-a, especially in the mid- and downstream during the wet season. This paper can advance NPSP research for small water bodies, elucidate the novel insights of nutrient "source-sink" dynamics, and provide a timely guide for rural planning and pond management in similar geo-climatic conditions. We recommend further studies on the structures and processes of MPSs to better understand similar NFWs: (1) incorporate landscape patterns, agronomic practices and ditch dynamics to better present extrinsic influences and their non-linear rainfall interactions; and (2) couple process-based models, embed smaller but non-negligible sources, and explore key environmental variables that trigger the "source-sink" role reversal.

References

Astel, A., Tsakovski, S., Barbieri, P., et al., 2007. "Comparison of self-organizing maps classification approach with cluster and principal components analysis for large environmental data sets." *Water Res.* 41(19): 4566-4578.

Bao, S. D., 2000. Soil agrochemical analysis. Beijing: China Agricultural Press.

Boardman, E., Danesh-Yazdi, M., Foufoula-Georgiou, E., et al., 2019. "Fertilizer, landscape features and climate regulate phosphorus retention and river export in diverse Midwestern watersheds." *Biogeochemistry* 146(3): 293-309.

Bolker, B. M., Brooks, M. E., Clark, C. J., et al., 2009. "Generalized linear mixed models: A practical guide for ecology and evolution." *Trends Ecol. Evol.* 24(3): 127-135.

Borrelli, P., Robinson, D. A., Fleischer, L. R., et al., 2017. "An assessment of the global impact of 21st century land use change on soil erosion." *Nat. Commun.* 8(1): 1-13.

Carlson, R. E., 1977. "A trophic state index for lakes." *Limnol. Oceanogr.* 22(2): 361-369.

Céréghino, R., and Park, Y. S., 2009. "Review of the self-organizing map (SOM) approach in water resources: Commentary." *Environ. Modell. Softw.* 24(8): 945-947.

Chen, L., Fu, B., Zhao, W., 2008. "Source-sink landscape theory and its ecological significance." *Frontiers of Biology in China.* 3(2): 131-136.

Chen, W., Duan, W., He, B., et al., 2017. "Water quality modeling for typical rural watershed based on the WASP model in Mountain Mao Region, upper Taihu Basin." *Journal of Lake Sciences.* 29(4): 836-847.

Chen, W., He, B., Nover, D.,et al., 2018. "Spatiotemporal patterns and source attribution of nitrogen pollution in a typical headwater agricultural watershed in Southeastern China." *Environ. Sci. Pollut. R.* 25(3): 2756-2773.

Chen, W., He, B., Nover, D., et al., 2019. "Farm ponds in southern China: Challenges and solutions for conserving a neglected wetland ecosystem." *Sci. Total Environ.* 659: 1322-1334.

Chen, W., Nover, D., Yen, H., et al., 2020. "Exploring the multiscale hydrologic regulation of multi-pond systems in a humid agricultural catchment." *Water Research.* 2020(184): 115987.1-115987.18

Cohen, M. J., Creed, I. F., Alexander, L., et al., 2016. "Do geographically isolated wetlands influence landscape functions?" *P. Natl. Acad. Sci. USA.* 113(8): 1978-1986.

Dai, L., 2019. "Implementing the water goals-The River Chief Mechanism in China." In *Politics and Governance in Water Pollution Prevention in China.* London: Palgrave Pivo.

Davies, D. L., and Bouldin, D. W., 1979. "A cluster separation measure." *IEEE Transactions on Pattern Analysis and Machine Intelligence* 1(2): 224-227.

Dirnböck, T., Kobler, J., Kraus, D., et al., 2016. "Impacts of management and climate change on nitrate leaching in a forested karst area." *J. Environ. Manage.* 165: 243-252.

Dollinger, J., Dagès, C., Bailly, J. S., et al., 2015. "Managing ditches for agroecological engineering of landscape. A review." *Agron. Sustain. Dev.*

35(3): 999-1020.

Dortch, Q., 1990. "The interaction between ammonium and nitrate uptake in phytoplankton." *Mar. Ecol. Prog. Ser.* 61(1): 183-201.

Fu, B., Merritt, W. S., Croke, B. F., et al., 2019. "A review of catchment-scale water quality and erosion models and a synthesis of future prospects." *Environ. Modell. Softw.* 114: 75-97.

Fu, B., Xu, P., Wang, Y., et al., 2018. "Assessment of the ecosystem services provided by ponds in hilly areas." *Sci. Total Environ.* 642: 979-987.

Golden, H. E., Creed, I. F., Ali, G., et al., 2017. "Integrating geographically isolated wetlands into land management decisions." *Front. Ecol. Environ.* 15(6): 319-327.

Golden, H. E., Rajib, A., Lane, C. R., et al., 2019. "Non-floodplain wetlands affect watershed nutrient dynamics: A critical review." *Environ. Sci. Technol.* 53(13): 7203-7214.

Hansen, A.T., Dolph, C.L., Foufoula-Georgiou, E., et al., 2018. "Contribution of wetlands to nitrate removal at the watershed scale." *Nat. Geosci.* 11(2): 127-132.

Haygarth, P. M., Jarvie, H. P., Powers, S. M., et al., 2014. "Sustainable phosphorus management and the need for a long-term perspective: The legacy hypothesis. Environ." *Sci. Technol.* 48(15): 8417-8419.

Jin, G., Xu, J., Mo, Y., et al., 2020. "Response of sediments and phosphorus to catchment characteristics and human activities under different rainfall patterns with Bayesian Networks." *J. Hydrol.* 584: 124695.

Jin, Y. H., Kawamura, A., Park, S. C., et al., 2011. "Spatiotemoral classification of environmental monitoring data in the Yeongsan River basin, Korea, using self-organizing maps." *J. Environ. Monit.* 13(10): 2886-2894.

Kanevski, M., Pozdnoukhov, A., Pozdnukhov, A., et al., 2009. *Machine learning for spatial environmental data: Theory, applications, and software.* Lausanne: EPFL Press.

Ki, S. J., Kang, J., Lee, S. W., et al., 2011. "Advancing assessment and design of stormwater monitoring programs using a self-organizing map: Characterization of trace metal concentration profiles in stormwater runoff." *Water Res.* 45(14): 4183-4197.

Kohonen, T., 1982. "A simple paradigm for the self-organized formation of structured feature maps." In *Competition and cooperation in neural nets,* edited by Amari, S. and Arbib, M.A. Berlin: Springer.

Kosten, S., Kamarainen, A. M. Y., Jeppesen, E., et al., 2009. "Climate-related differences in the dominance of submerged macrophytes in shallow lakes." *Global Change Biol.* 15(10): 2503-2517.

Laird, N.M., and Ware, J.H., 1982. "Random effects models for longitudinal data." *Biometrics* 38: 963-974.

Lane, C. R., Leibowitz, S. G., Autrey, B. C., et al., 2018. "Hydrological, physical, and chemical functions and connectivity of non-floodplain wetlands to downstream waters: A review." *J. Am. Water Resour. As.* 54(2): 346-371.

Lessels, J. S., and Bishop, T. F. A., 2013. "Estimating water quality using linear mixed models with stream discharge and turbidity." *J. Hydrol.* 498: 13-22.

Li, D., Zheng, B., Chu, Z., et al., 2019. "Seasonal variations of performance and operation in field-scale storing multi-pond constructed wetlands for nonpoint source pollution mitigation in a plateau lake basin." *Bioresource Technol.* 280: 295-302.

Li, F., Cheng, S., Yu, H., et al., 2016. "Waste from livestock and poultry breeding and its potential assessment of biogas energy in rural China." *J. Clean. Prod.* 126: 451- 460.

Lischeid, G., Kalettka, T., Holländer, M., et al., 2018. "Natural ponds in an agricultural landscape: External drivers, internal processes, and the role of the terrestrial-aquatic interface." *Limnologica* 68: 5-16.

Liu, Y., Fu, Q., Yin, C., 2009. "Phosphorus sorption and sedimentation in a multi-pond system within a headstream agricultural watershed." *Water Qual. Res. J. Can.* 44(3): 243-252.

Luo, P., Kang, S., Apip, Z. M., et al., 2019. "Water quality trend assessment in

Jakarta: A rapidly growing Asian megacity." *PLoS One* 14(7): e0219009.

Marton, J. M., Creed, I. F., Lewis, D. B., et al., 2015. "Geographically isolated wetlands are important biogeochemical reactors on the landscape." *Bioscience* 65(4): 408-418.

MEP (Ministry of Environmental Protection of China), 2002. *Environmental Quality Standards for Surface Water (GB 3838-2002)*. Beijing: China Environmental Science Press.

Milošević, D., Čerba, D., Szekeres, J., et al., 2016. "Artificial neural networks as an indicator search engine: The visualization of natural and man-caused taxa variability." *Ecol. Indic.* 61(2): 777-789.

Mitsuo, Y., Tsunoda, H., Kozawa, G., et al., 2014. "Response of the fish assemblage structure in a small farm pond to management dredging operations." *Agr. Ecosyst. Environ.* 188: 93-96.

Nguyen, T. T., Kawamura, A., Tong, T. N., et al., 2015. "Clustering spatio-seasonal hydrogeochemical data using self-organizing maps for groundwater quality assessment in the Red River Delta. Vietnam." *J. Hydrol.* 522: 661-673.

Nie, X., Li, H., Li, X., 2012. "Comparison of nitrogen and phosphorus removal efficiencies by storm runoffs for the ponds in the upper and lower reaches of a typical sub-catchment in Lake Chaohu drainage basin." *Journal of Lake Sciences* 24(1): 89-95.

Nosakhare, U. H., and Bright, A. F., 2017. "Evaluation of techniques for Univariate Normality Test using Monte Carlo simulation." *American Journal of Theoretical and Applied Statistics* 6(5-1): 51-61.

Ouyang, W., Guo, B., Cai, G., et al., 2015. "The washing effect of precipitation on particulate matter and the pollution dynamics of rainwater in downtown Beijing." *Sci. Total Environ.* 505: 306-314.

Park, Y. S., Kwon, Y. S., Hwang, S. J., et al., 2014. "Characterizing effects of landscape and morphometric factors on water quality of reservoirs using a self-organizing map." *Environ. Modell. Softw.* 55: 214-221.

Pinheiro, J. C., and Bates, D. M., 2000. "Linear mixed-effects models: Basic concepts and examples." *In Mixed-effects models in S and S-Plus,* New York: Springer.

Poschlod, P., and Braun-Reichert, R., 2017. "Small natural features with large ecological roles in ancient agricultural landscapes of Central Europe-History, value, status, and conservation." *Biol. Conserv.* 211: 60-68.

Qiao, X., Du, J., Kota, S. H., et al., 2018. "Wet deposition of sulfur and nitrogen in Jiuzhaigou National Nature Reserve, Sichuan, China during 2015-2016: Possible effects from regional emission reduction and local tourist activities." *Environ. Pollut.* 233: 267-277.

Qin, B., Zhou, J., Elser, J. J., et al., 2020. "Water depth underpins the relative roles and fates of nitrogen and phosphorus in lakes." *Environ. Sci. Technol.* 54(6): 3191-3198.

Romić, D., Castrignanò, A., Romić, M., et al., 2020. "Modelling spatial and temporal variability of water quality from different monitoring stations using mixed effects model theory." *Sci. Total Environ.* 704: 135875.

Scott, D. T., Gomez-Velez, J. D., Jones, C. N., et al., 2019. "Floodplain inundation spectrum across the United States." *Nat. Commun.* 10(1): 1-8.

SEPA (State Environmental Protection Administration of China), 2002. *Standard methods for water and wastewater monitoring and analysis, 4th ed.* Beijing: China Environmental Science Press.

Shen, Z. Y., Hou, X., Li, W., Aini, G., 2014. "Relating landscape characteristics to nonpoint source pollution in a typical urbanized watershed in the municipality of Beijing." *Landscape and Urban Planning* 123: 96-107.

Sønderup, M. J., Egemose, S., Hansen, A. S., et al., 2016. "Factors affecting retention of nutrients and organic matter in stormwater ponds." *Ecohydrology* 9(5): 796-806.

Thornhill, I. A., Biggs, J., Hill, M. J., et al., 2018. "The functional response and resilience in small waterbodies along land-use and environmental gradients." *Global Change Biol.* 24(7): 3079-3092.

Thorslund, J., Jarsjö, J., Jaramillo, F., et al., 2017. "Wetlands as large-scale nature-based solutions: Status and challenges for research, engineering and management." *Ecol. Eng.* 108: 489-497.

Tu, J., 2013. "Spatial variations in the relationships between land use and water quality across an urbanization gradient in the watersheds of northern Georgia, USA." *Environ. Manage.* 51(1): 1-17.

UN-Water (The United Nations World Water Assessment Programme), 2018. The United Nations World Water Development Report 2018: Nature-Based Solutions for Water. Paris, UNESCO.

Usio, N., Nakagawa, M., Aoki, T., et al., 2017. "Effects of land use on trophic states and multi-taxonomic diversity in Japanese farm ponds." *Agr. Ecosyst. Environ.* 247: 205-215.

Verhoeven, J. T., Arheimer, B., Yin, C., et al., 2006. "Regional and global concerns over wetlands and water quality." *Trends Ecol. Evol.* 21(2): 96-103.

Vesanto, J., and Alhoniemi, R., 2000. "Clustering of the self-organizing map." *IEEE Trans. Neural Netw.* 11(3): 586-600.

Wang, X., and Zhang, F., 2018. "Multi-scale analysis of the relationship between landscape patterns and a water quality index (WQI) based on a stepwise linear regression (SLR) and geographically weighted regression (GWR) in the Ebinur Lake oasis." *Environ. Sci. Pollut. R.* 25(7): 7033-7048.

Williams, P., Biggs, J., Crowe, A., et al., 2010. *Countryside survey: Ponds report from 2007.* Technical report No.7/07 Pond Conservation and NERC/ Centre for Ecology & Hydrology,77pp. (CEH Project Number:C03259).

Williams, R., and Fryirs, K., 2016. "Management and conservation of a unique and diverse Australian river type: Chain-of-ponds." The 8th Australian Stream Management Conference, River Basin Management Society, Australia.

Xia, Y., Ti, C., She, D., et al., 2016. "Linking river nutrient concentrations to land use and rainfall in a paddy agriculture-urban area gradient watershed in southeast China." *Sci. Total Environ.* 566-567: 1094-1105.

Xu, W., Fan, X., Ma, J., et al., 2019. "Hidden loss of wetlands in China." *Curr.*

Biol. 29(18): 3065-3071.

Yan, P., Shen, C., Fan, L., et al., 2018. "Tea planting affects soil acidification and nitrogen and phosphorus distribution in soil." *Agr. Ecosyst. Environ.* 254: 20-25.

Yanai, R. D., Fahey, T. J., Miller, S. L., 1995. *Efficiency of nutrient acquisition by fine roots and mycorrhizae. Resource physiology of conifers.* Florida: Academic Press.

Yen, H., White, M. J., Arnold, J. G., et al., 2016. "Western Lake Erie Basin: Soft-data-constrained, NHDPlus resolution watershed modeling and exploration of applicable conservation scenarios." *Sci. Total Environ.* 569: 1265-1281.

Yu, K. J., Jiang, Q. Z., Wang, Z. F., et al., 2015. "The research progress and prospect of Beitang landscape." *Areal Research and Development* 34(3): 130-136.

Zhang, W., Li, H., Pueppke, S. G., et al., 2020a. "Nutrient loss is sensitive to land cover changes and slope gradients of agricultural hillsides: Evidence from four contrasting pond systems in a hilly catchment." *Agri. Water Manage.* 237: 106165.

Zhang, X., Zhi, X., Chen, L., et al., 2020b. "Spatiotemporal variability and key influencing factors of river fecal coliform within a typical complex watershed." *Water Res.* 178: 115835.

Zhang, Y., Zhou, M., Wu, X., 1986. Soil records of Jurong Country in Jiangsu Province. Soil Survey Office of Jurong County, Zhenjiang Department of Agriculture.

Zou, L., Liu, Y., Wang, Y., et al., 2020. "Assessment and analysis of agricultural non-point source pollution loads in China: 1978-2017." *J. Environ. Manage.* 263(5): 110400.

Appendix

Text A5.1

Biomass was prepared in the following procedures: (1) collect macrophyte samples in several 0.5m^2 quadrats, which represented the single or combination of species in the pond; (2) dry the sample tissues to a constant weight (see Bowden et al., 2017 for detailed procedures); and (3) reckon the area-weighted average of these samples, in which the area-weight corresponded to the single or combination of species that each sample represented. Note that the resulting dry weight was duplicated for field surveys in the next year, owing to the relatively stable species richness and distribution in each pond.

Text A5.2

With detailed computation procedures in Anselin (2003) and Chang (2008), the spatial autocorrelation was calculated based on all sampling sites, while the temporal autocorrelation was based on 12 months of the year. The weight matrix in the calculation was prepared via a [0, 1] binary adjacency as suggested by Getis, 2009. With values between-1 and 1, a Moran's I of-1, 1, and 0 means perfect negative and positive autocorrelation, as well as randomness among the sampling sites or periods (Moran et al., 1950; Tillé et al., 2018). The global autocorrelation, SOM, and LMEM analysis were performed using the Spatial Analysis toolbar of GeoDa 1.14, Neural Network Clustering application of Matlab R2018b, and Mixed-effects Model toolbox of SPSS 26, respectively.

Text A5.3

In Eq. (1), Concentration corresponds to the water quality variables. Intercept is a constant representing the background value of each dependent variable. The random error ε is the variance, which is assumed to be independent and identically distributed with a mean value equal to 0. The first 8 vector operations represent the dot product of each variable in the category and its corresponding coefficient. For example, b·Landuse represents $b_{Water} \times$

%Water + b_{Forest} × %Forest + $b_{Agricultural\_land}$ × %Agricultural_land + $b_{Rural\_residential\_area}$ × %Rural_residential_area, in which % means the area proportion of each landuse type in the drainage area. In the last 7 vector operations, rainfall is first multiplied by each variable, and then taken dot product with the corresponding coefficient. For example, B·(Landuse × rainfall) represents B_{Water} × %Water × Rainfall + B_{Forest} × %Forest × Rainfall + $B_{Agricultural\_land}$ × %Agricultural_land × Rainfall + $B_{Rural\_residential\_area}$ × %Rural_residential_area × Rainfall, in which rainfall is the monthly cumulative value. Hence, there is 1 coefficient in a, 4 in b, 1 in c, 1 in d, 1 in e, 3 in f, 3 in g, 1 in h, and totally 29 in Eq. (1).

Text A5.4

Within the framework of null hypothesis significance testing (NHST, Johnson et al., 2015), and using the SIMR software package, the power analysis aims to estimate the minimum sample size required by a designed regression model to achieve an acceptable significance level (typically $p < 0.05$) and power level (see Green and MacLeod, 2016 for detailed theories, and a value of 0.80 is conventionally adequate according to Field et al., 2007).

Text A5.5

The synergistic influences with monthly cumulative rainfall explained ~ 76, 79, 68, 77, 60, and 76% of the NH_4^+-N, NO_3^--N, TN, TP, COD_{Mn}, and Chl-a variations for the upstream ponds in the wet season, respectively (Fig. A5.4). The degrees of explanation were ~ 83, 72, 62, 75, 76, and 59% in the dry season, respectively, with an average less than that of the wet season (~ 71 vs. 73%). The midstream ponds had better modelling performance for TN, TP, COD_{Mn}, and Chl-a in the wet season (~ 72, 81, 62, and 80%, respectively), and for NO_3^--N, TN, and Chl-a in the dry season (~ 79, 76, and 73%, respectively) (Fig. A5.5). Similarly, the downstream ponds were only better explained for NO_3^--N and COD_{Mn} variations in the wet season (~ 73 and 70%, respectively), and Chl-a changes in the dry season (~ 79%), compared to the midstream simulations (Fig. A5.6). The reduced performance downgradient was presumably due to the intermittent, uncertain pollution discharge from poultry feeding and rural households in the mid- and downstream (Xia et al., 2016; Chen et al., 2018; Fu et al., 2018). Moreover, the average performance of the wet season was gradually lower than that of the dry season in the mid- and downstream ($\sim$ 72 vs 74%, and 66 vs. 71%, respectively), probably due to the increasingly complex, interlaced ditch networks in agricultural land, whose hydrologic and nutrient retention are more functional during rainfall-runoff processes (Li et al., 2019; Chen et al., 2020). According to the Kolmogorov-Smirnov One Sample Test (Berger and Zhou, 2014) and corresponding two-tailed P-value results (α = 0.05), the residuals of the 36 models were all normally distributed, which again demonstrated the effectiveness of LMEM simulations. Besides, grouped diagnostics indicated that the range of residuals enlarged slightly as the water quality observations increased, but sometimes narrowed at last (e.g., TP in the upstream, and TN, COD_{Mn} and Chl-a in the midstream), which was largely due to the limited number of samples in the high-value regions.

Text A5.6

The abbreviations include pond extrinsic factors: (1) area proportion of agricultural land (%Agri), forest (%Forest), rural residential areas (%Resi), and water (%Water), (2) area proportion of Dystric Regosols (%Rego), (3) number of chickens and ducks (#Poultry); and intrinsic factors: (4) TN (*TN*) and TP (*TP*) concentrations of surface sediments, (5) biomass (Biomass) and percentage cover (%Cover) of macrophytes; as well as (6) monthly cumulative rainfall (Rain) and its linear effect on the above variables (× Rain). Coefficients that reduced modelling performance from up- to downstream were not shown.

Table A5.1 Water quality, sediment, and macrophyte variables across the 2.5-year experimental period. Note: The Chl-a concentrations, sediment and macrophyte variables were measured bimonthly, seasonally (i.e., wet and dry) and trimonthly, respectively, due to their relatively lower variability.

Category	Variable (unit)	Period	Sample size	Category	Variable (unit)	Period	Sample size
Water quality	NH_4^+-N (mg/L)	Monthly	690	Sediment	TN (mg/kg)	Seasonally	138
	NO_3^--N (mg/L)	Monthly	690		TP (mg/kg)	Seasonally	138
	TN (mg/L)	Monthly	690		TOC (%)	Seasonally	138
	TP (mg/L)	Monthly	690	Macrophytes	Species richness	Trimonthly	230
	COD_{Mn} (mg/L)	Monthly	690		Percentage cover (%)	Trimonthly	230
	Chl-a (µg/L)	Bimonthly	345		Biomass (g/m²)	Trimonthly	230

Table A5.2 Global Moran's *I* values of the autocorrelation analysis

	Spatial autocorrelation	Temporal autocorrelation
NH_4^+-N	0.31	0.14
NO_3^--N	-0.12	0.41
TN	0.10	0.15
TP	0.34	-0.12
COD_{Mn}	0.07	-0.31
Chl-a	0.23	0.27

Table A5.3 Response variables that log transformed to meet normal distribution assumptions

Response variable	Upstream		Midstream		Downstream	
	Wet season	Dry season	Wet season	Dry season	Wet season	Dry season
NH_4^+-N	Yes	No	No	No	No	Yes
NO_3^--N	No	No	No	No	No	No
TN	No	No	No	No	No	No
TP	Yes	Yes	No	Yes	Yes	Yes
COD_{Mn}	No	No	No	Yes	No	Yes
Chl-a	No	No	Yes	No	No	No

Fig. A5.1 (a) DEM, (b) land use, (c) soil classes, and (d) free-range poultry feeding of the MPS catchments.

Fig. A5.2 SOM planes of the 6 water quality variables for all sampling sites and periods

Note: The total number of neuron nodes was 80, with a row number of 10 and a column number of 8. In the feature planes, NO_3^--N, TN, and TP increased from the lower to upper left part of the plane, indicating similar spatiotemporal patterns of these variables. COD_{Mn} and Chl-a, however, present a totally different trend, which increased from the upper right to the lower left area. In-between the above patterns, NH_4^+-N increased from right to the mid-left part.

Fig. A5.3 Clustering water quality observations: (a) selection of the optimal cluster number; (b) dendrogram of the hierarchical clustering; and (c) cluster boundaries of the SOM neuron nodes that represent all water quality variables.

Note: The optimal cluster number, 6, in (a) was determined by the partitional clustering and lowest value of Davies-Bouldin index. SOM neuron ID in (b) corresponded to the above cluster ID in the same color. The boundary of each cluster in (c) was delineated by the hierarchical analysis and Ward's linkage criterion, while the label of each neuron was the total number of water samples linked to that neuron.

Fig. A5.4 The LMEM plots and performance for the upstream ponds across the 6 water quality variables and 2 seasons

Note: The blue sample triangles and fitting lines belong to the wet season, while the yellow ones belong to the dry season. Residuals diagnostics was presented at the bottom right corner. The two-tailed P-values ($\alpha=0.05$) was calculated via the Kolmogorov-Smirnov One Sample Test (Berger and Zhou, 2014), while the grouped residuals against observations follow the same color pattern. The NH_4^+-N levels in the wet season and TP levels in both seasons were log transformed to meet normal distribution assumptions of the response variables.

Fig. A5.5 The midstream LMEM plots and performance. Notes are the same to Fig. A5.4, while the TP and COD_{Mn} levels in the dry season and Chl-a levels in the wet season were log transformed.

Fig. A5.6 The downstream LMEM plots and performance. Notes are the same to Fig. A5.4, while the NH_4^+-N, TP and COD_{Mn} levels in the dry season and TP levels in the wet season were log transformed.

Appendix references

Anselin, L., 2003. GeoDa 0.9 user's guide. Department of Agricultural and Consumer Economics, University of Illinois, IL, USA.

Berger, V. W., and Zhou, Y., 2014. Kolmogorov-Smirnov test: Overview. Wiley StatsRef: Statistics Reference Online.

Bowden, W. B., Glime, J. M., Riis, T., 2017. *Macrophytes and bryophytes. Methods in Stream Ecology.* Florida: Academic Press.

Chang, H, 2008. "Spatial analysis of water quality trends in the Han River basin, South Korea." *Water Res.* 42(13): 3285-3304.

Chen, W., He, B., Nover, D., et al., 2018. "Spatiotemporal patterns and source attribution of nitrogen pollution in a typical headwater agricultural watershed in Southeastern China. Environ." *Sci. Pollut. R.* 25(3): 2756-2773.

Chen, W., Nover, D., Yen, H., et al., 2020. "Exploring the multiscale hydrologic regulation of multi-pond systems in a humid agricultural catchment." *Water Res.* 1847:115987.

Field, S. A., O'Connor, P. J., Tyre, A. J., et al., 2007. "Making monitoring meaningful." *Austral Ecol.* 32(5): 485-491.

Fu, B., Xu, P., Wang, Y., et al., 2018. "Assessment of the ecosystem services provided by ponds in hilly areas." *Sci. Total Environ.* 642: 979-987.

Getis, A., 2009. "Spatial weights matrices." *Geographical Analysis* 41(4): 404-410.

Green, P., and MacLeod, C. J., 2016. "SIMR: An R package for power analysis of generalized linear mixed models by simulation." *Methods Ecol. Evol.* 7(4): 493-498.

Johnson, P. C., Barry, S. J., Ferguson, H. M., et al., 2015. "Power analysis for generalized linear mixed models in ecology and evolution." *Methods Ecol. Evol.* 6(2): 133-142.

Li, D., Zheng, B., Chu, Z., et al., 2019. "Seasonal variations of performance and operation in field-scale storing multi-pond constructed wetlands for nonpoint source pollution mitigation in a plateau lake basin." *Bioresource*

Technol. 280: 295-302.

Moran, P. A., 1950. "Notes on continuous stochastic phenomena." *Biometrika* 37: 17-23.

Tillé, Y., Dickson, M. M., Espa, G., et al., 2018. "Measuring the spatial balance of a sample: A new measure based on Moran's I index." *Spatial Statistics* 23: 182-192.

Xia, Y., Ti, C., She, D., et al., 2016. "Linking river nutrient concentrations to land use and rainfall in a paddy agriculture-urban area gradient watershed in southeast China." *Sci. Total Environ.*566-567: 1094-1105.

Chapter
⑥

Syndicating Multi-source Data for Inundation Extent Analysis of Small Reservoirs

6.1 Introduction

During the past 100 years, more than 847,000 reservoirs have been constructed globally, and approximately 95% of these are small reservoirs, whose heights are less than 15 m from foundation to crest (Rosenberg et al., 2000; World Commission on Dams, 2000). They provide water for multiple uses, including drinking, hydroelectric energy and recreation, and stabilize extreme flows to mitigate floods or droughts. However, dam failures with potentially catastrophic consequences for downstream communities and the environment are set to increase due to climatic and hydrological extremes (e.g., heat waves and heavy precipitation events), deterioration of older infrastructure, and mismanagement (Hirabayashi et al., 2013; Pisaniello et al., 2015; Döll et al., 2015; Giorgio et al., 2017). Among the structural and non-structural measures to mitigate flood risks, inundation extent analysis is one of the most important components, as it extends information that is helpful in dealing with flood early warning, flood control operations, and hydraulic structure design (Sarhadi et al., 2012; Shen et al., 2015; Teng et al., 2017). Although inundation simulations have been made possible in river channels, urban areas, and floodplains (Sanders et al., 2010; Yamazaki et al., 2012; Demir et al., 2013; Liu et al., 2016), quick and effective estimation of submerged extent in and around small reservoirs, especially on associated hydraulic structures, such as embankment dams, are less studied, and garnering increased attention from hydrologists, civil engineers, and decision-makers (Pisaniello et al., 2015; Chen et al., 2017; Sampson et al., 2012).

As a prerequisite to inundation extent analysis, terrain modeling based on various kinds of Digital Elevation Models (DEMs) has been widely used. Satellite-derived datasets, such as SRTM DEM and ASTER GDEM, have enormous utility as they cover almost the entire earth, although their spatial resolution is often too coarse for scales of interest for hydrologic applications (Yamazaki et al., 2012). Recent advances in remote sensing technologies, such as airborne and terrestrial Light Detection and Ranging (LiDAR), have enabled the collection of topographic data with a high spatial resolution and vertical accuracy (Sampson et al., 2012; Andersen et al., 2017). However, data accessibility is an essential consideration regarding numerous and widely distributed small reservoirs. Available datasets include the 5-m resolution DEM produced by China's National Spatial Data Infrastructure, which covers the central and eastern provinces (Chen et al., 2014), and the 3 to 5 m resolution DEMs published by the National Digital Elevation Program, which is available for 43 percent of the Continental U.S. (Snyder et al., 2012). These airborne LiDAR-derived DEMs are post-processed to represent bare-earth elevations, and easy to access from government bureaus, but have two major flaws, which may impede reservoir inundation extent analysis at very high and low water stages. One is the unclear imaging of associated hydraulic structures (Turner et al., 2013; Zhang et al., 2014), and the other is the lack of reservoir bathymetry (Pan et al., 2013).

Multi-source and heterogeneous spatial data are usually assembled to create realistic and coherent terrain models for inundation simulations. Researchers have addressed questions such as integrating river cross-section lines with surrounding floodplain topography for 2D/3D hydrodynamic modeling (Cook et al., 2009; Conner et al., 2014), setting up building blocks or vertical walls to represent man-made structures according to the aerial imagery and static properties for urban flood routing (Turner et al., 2013; Chen et al.,

2012; Li et al., 2014), and using pre-defined drainage networks to remove pits in DEMs caused by vegetation canopies, sub-pixel sized structures, and random radar speckles (Yamazaki et al., 2012). These studies have shown that the effectiveness of inundation extent analysis at meso- and micro-scales can be improved by the combined use of DEMs and other topographic datasets. Following the trend in multi-source and heterogeneous data integration, we advocate the use of two common tools to improve terrain modeling, and to represent extreme water stages in small reservoirs. These tools include bathymetric measurements for reservoir stage-storage relationship, and 3D geometric models for hydraulic structure design.

Water stage is an essential factor in inundation extent analysis because submerged extent is determined by comparing the elevation of each terrain mesh cell with the corresponding water stage at a given time. There are essentially three methods for obtaining such information described below. Spatial interpolation of on-site observations is quite efficient, but requires significant advance investment in hydrological stations and gauging systems (Pan et al., 2013; Lay-Ekuakille et al., 2017; Manian et al., 2012). Although 2D/3D hydrodynamic models are generally used for flood routing simulations, their iterative solutions require sophisticated computing resources, such as parallel computer architecture, to generate real-time risk analysis (Sanders et al., 2010; Liu et al., 2016; Sampson et al., 2012; Yadav et al., 2016; Vacondio et al., 2014). Alternatively, another simplified approach uses a horizontal water-surface, which is obtained by the actual water-storage and stage-storage relationship, to estimate the submerged extent. This method has been widely adopted in standard GIS software, such as the ArcGIS Hydrology toolset, and the Crayfish Plugin for Quantum GIS, and has achieved satisfactory results in urban storm-inundation simulations (Neubert et al., 2014; Zhang et al., 2014; CH2M HILL, 2017), loss assessment of river levee breaches (Zhang et al., 2014; Krupka et al., 2009), and determination of the stage-area relationship in lakes and wetlands (Pan et al., 2013; Chen et al., 2017). Although the flat-water surface ignores hydraulic gradients in flowing floods, it has the advantage of a fast calculation speed, and is sufficiently robust to support flood control management in small reservoirs.

Based on terrain models and water stage measurements, inundation extent is usually delineated by two specific algorithms, which are non-source flooding and source flooding, according to whether the estimation considers flow connectivity (Shen et al., 2015; Zhang et al., 2014; Noman et al., 2001). Non-source flooding is equivalent to a large area receiving heavy and uniform precipitation, with all terrain mesh cells below the water surface belonging to the submerged area. In contrast, source flooding compares the water stage with terrain elevations and considers flow continuity, so inundation can only occur in places where water flows. Both non-source flooding and source flooding rely on a pre-established neighboring relationship of the terrain mesh cells to spread submerged areas. The relationship is innate in DEM data structures, but absent in 3D geometric models. Therefore, the 3D geometric models are mostly treated as "transparent" or "isolated" from the two inundation algorithms (Sarhadi et al., 2012; Li et al., 2014; Zhang et al., 2013), although they can represent topographic details in relatively coarse-resolution DEMs, making it unrealistic to delineate submerged extent on associated hydraulic structures in small reservoirs.

In view of the above scientific challenges, this study aims to develop a method to achieve rapid and effective estimation of inundation extent for both small reservoirs and associated hydraulic structures. To improve the effectiveness of inundation extent analysis, especially at very high and low water stages, topography, bathymetry and a 3D dam model are integrated to create a realistic and coherent reservoir terrain surface. To efficiently delineate inundation extent on both the natural terrain and the 3D dam model, the source flooding algorithm based on a horizontal water surface is enhanced by introducing the intersection detection from computational geometry, so that submerged areas can spread from terrain mesh cells to a 3D geometric model.

Taking the Hengshan Reservoir in Southeastern China as the study area, the proposed methods are tested by a historical rainfall event during the typhoon season, and validated by a comparison between the derived and existing stage-storage relationships. As a first attempt to involve 3D geometric models of hydraulic structures into inundation extent analysis, this study can inform other flood simulations and risk assessments at meso- and micro-scales for hydrologists. They are also useful for geospatial analysis and geo-computation that involve both natural terrains and man-made structures for GIS specialists and decision-makers.

6.2 Methodology

6.2.1 Study area and materials

Along the boundary of Wuxi City and Changzhou City in Southeastern China, the Hengshan Reservoir lies in the subtropical monsoon climate zone, with the mean annual precipitation of 1310 mm, and the mean annual temperature of 15.7 (Fig. 6.1). There were three homogeneous earth-filled dams dating from the reservoir construction project in the 1960s: the main embankment dam, and the east and west auxiliary dams. After operating for more than 30 years, another reinforcement project was implemented during 2001—2004. The current water storage is from 50 to 80 million m^3, and the existing stage-storage relationship for reservoir daily management is established based on bathymetric measurements below the water surface and a polynomial extrapolation for the surrounding natural terrain and embankment dams.

Fig. 6.1 Study area of the Hengshan Reservoir

Fig. 6.2 shows the preparations of topographic and bathymetric datasets for reservoir terrain modeling. A 5-m resolution airborne LiDAR-derived DEM is obtained from the Jiangsu Provincial Bureau of Surveying and Mapping in China (Fig. 6.2A). These show water surface elevations, rather than the bathymetric topography in the water areas, so 2,740 bathymetry points with an average interval of 61.3 m in the center, and 12.3 m in riparian areas are also collected using a boat-mounted sonar device in conjunction with the global positioning system (Fig. 6.2B). In addition, the 2-m resolution optical imagery is obtained from Google Earth Pro Version, and georeferenced as supplementary data to help identify geographic features (Fig. 6.2C).

Fig. 6.2 The DEM (A), bathymetric measurements (B), and Google imagery (C) for reservoir terrain modeling.

The 3D dam model created by 3DS Max is analyzed for terrain integration and inundation extent analysis (Fig. 6.3). The model is partially created on the natural terrain surface, with possible discrepancies from the DEM data at that location. Different structural components are deliberately inserted into each other, instead of snapping to the vertices and edges, making it easier to generate the realistic model appearance, but introducing significant false intersections (Fig. 6.3B). Small defects, such as missing components, are inevitable in this synthetic analytical approach, resulting in unexpected cracks on the model surface (Fig. 6.3C).

Fig. 6.3 The 3D dam model and its structural characteristics. The panorama and local close-up are in (A), while the issues of intersected components and unexpected small defects are in (B) and (C).

6.2.2 Integrating bathymetry with surrounding topography

Geometric descriptions of the reservoir bathymetry and surrounding topography serve as a prerequisite to inundation extent analysis (Cook et al., 2009; Saksena et al., 2015; Curtarelli et al., 2015). Although the emerging technology of airborne topo-bathymetric LiDAR can provide rapid collection of seamless, high-resolution terrain information on both sides of the land-water interface, it requires several resource-intensive procedures, including data filtering, water surface detection, and refraction correction, to create a ready-to-use DEM product (Andersen et al., 2017; Yadav et al., 2016). Another common and practical method is to create coherent terrain models by integrating bathymetric measurements with wide-coverage, but relatively coarse-resolution DEMs. Existing literature focuses on the overlap between these disparate datasets, due to different timestamps or spatial scales, and recommends the combined use of buffer polygons and weighted linear interpolation to accurately represent terrain elevations in riparian areas (Conner et al., 2014; Curtarelli et al., 2015). For example, the elevation of the buffer boundary at the river levee receives a higher weight than the elevation of the buffer boundary in the surrounding floodplain, when the interpolating point is closer to the river levee. However, this semi-automated method relies heavily on field surveys to set up the buffer polygons and related weights, and ignores additional datasets such as satellite imagery to facilitate the terrain integration by identifying geographic features that are not distinctly defined by topographic data.

Considering the limitations of existing techniques, an auxiliary method employing contour lines and Google imagery is proposed to integrate bathymetric measurements with airborne LiDAR-derived DEMs, so that a continuous reservoir terrain surface can be prepared for inundation extent analysis to get elevations for finite element meshes. First, contour lines of the bathymetry points and DEM data are repeatedly generated at different intervals until the geographic features, such as floodwalls, floodplains, islands, and water areas, are roughly delineated on the imagery. Second, contour lines, bathymetry points, and topographic points of the raster DEM are incorporated, and the reservoir area is manually divided into four regions: land, water, and riparian with and without bathymetric measurements (Fig. 6.4). In each region, the disparate elevation point sets are merged according to the following principles:

Fig. 6.4 The subregions for elevation point integration, where (A) is land, (B) stands for water, and (C) and (D) are riparian with and without bathymetric measurements, respectively.

(1) Land areas, including floodwalls, floodplains and islands, are delineated based on optical information in the imagery and specific topographic contour lines on the land-water interface (Fig. 6.4A), where only topographic points are represented.

(2) Similarly, water is defined as being surrounded by bathymetric contour lines nearest to the floodwalls or floodplains, and bathymetric contour lines closest to the islands (Fig. 6.4B), where only bathymetry points remain.

(3) The areas between land and water boundaries are manually divided into several small areas, according to the density of bathymetry points. The areas with high point-density are covered by the bathymetric surveys (Fig. 6.4C), so bathymetry points, which undergo fewer post-processing steps than the DEM data, are maintained.

(4) Small areas with scarce or no bathymetry points may have been land areas when the surveys were preformed, but they are ambiguously regarded as water according to the land boundaries (Fig. 6.4D). The buffer polygons and weighted linear interpolation are established only in these places to eliminate abrupt elevation changes. According to Cook and Merwade, (2009), the same elevation weight is given to interpolating points between water and the surrounding floodplains, whereas higher weights of dam elevations are given to

points which are closer to the embankment dams.

A Delaunay triangulation of the merged point set is constructed based on a fast algorithm with linear time-complexity, whose main steps, including segmentation, generation and insertion, were elaborated in Janardan et al. (Janardan et al., 2003). The terrain meshes are organized in a matrix data format, instead of object-oriented data structures (Fig. 6.5), so that matrix operations are possible, including basic arithmetic, inversion, transposition, and quick search (Walt et al., 2011). Data matrices consist of a vertex matrix, an edge matrix, and a triangle matrix with properties and examples in Fig. 6.5B. For the sake of consistency, vertex IDs are saved in ascending order for each edge in the edge matrix, and in a clockwise direction for each triangle in the triangle matrix, so the edge IDs and neighboring triangle IDs are determined by the corresponding vertices and edges, respectively. In addition, the neighboring triangle IDs are marked as "null" for the edges on the border of terrain meshes.

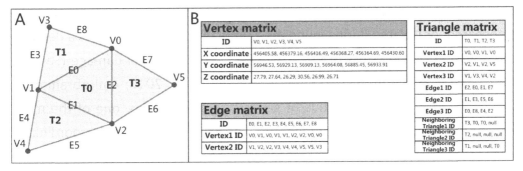

Fig. 6.5 The Delaunay triangulation for reservoir terrain modeling. An example of the terrain meshes is in (A), and the corresponding data matrices are in (B).

6.2.3 Integrating natural terrain with a 3D dam model

Floods and inundation occur naturally but can also be affected by human engineering activities. Accurate representation of finer scale topographic features, such as buildings, bridges, and embankment dams, is therefore essential for inundation extent analysis and damage assessment at meso- and micro-scales (Turner et al., 2013; Selvanathan et al., 2010; Amirebrahimi et al., 2016), and has proven useful to improve public participation in the design and construction of hydraulic structures (Lai et al., 2011). Researchers have proposed several methods to represent man-made structures during inundation simulations, among which the most prevalent techniques are a) establishing building blocks based on static properties, such as locations and heights (Turner et al., 2013; Chen et al., 2012; Li et al., 2014); b) employing vector lines or constrained meshes to set up vertical walls of water retaining structures, e.g., river levees (Shen et al., 2015; Zhang et al., 2014); and c) incorporating external 3D geometric models to fully simulate the appearance of a man-made structure (Zhang et al., 2013; Lai et al., 2011; Biljecki et al., 2015). Although the first two approaches were validated by laboratory experiments in the IMPACT project (Alcrudo, 2004), they treat man-made structures as 1D point-like and 2D linear features, where extra settings are required for their appearance. In contrast, 3D geometric models with inherent visualization advantages are increasingly being employed in hydrologic applications and provide important information related to flood risk assessments. However, these models are mostly regarded as transparent or isolated during inundation simulations, which impedes a detailed investigation of the effect of water blockage and structural damage (Sarhadi et al., 2012; Zhang et al., 2013; Amirebrahimi et al., 2016; Amirebrahimi et al., 2016). From the perspective of geometric descriptions, one of the main reasons is that 3D geometric models are only placed at appropriate coordinates in the DEM data, instead of being seamlessly integrated into the natural terrain surface.

We present a 3D dam model, created from construction drawings, and integrated seamlessly into natural terrain data, so that the simplifications and distortions related to the morphology of associated hydraulic structures are eliminated, and inundation extent can cover the entire reservoir area with a detailed representation of water stage on embankment dams. After placing the 3D dam model in the right position in the DEM, one can see that the bottom has covered several terrain mesh cells, requiring delineation of a suitable blending boundary and seamless replacement of the terrain cells below. The detailed procedures are as follows:

(1) Convert the 3D dam model from .3DS (a native file format in 3DS Max) to. OBJ (an open file format for 3D geometric modeling). As in the case of the terrain model, it is organized as surface triangular meshes (Fig. 6.6A). The blending boundary is then manually extracted along the edge vertices in a counterclockwise direction (Fig. 6.6B).

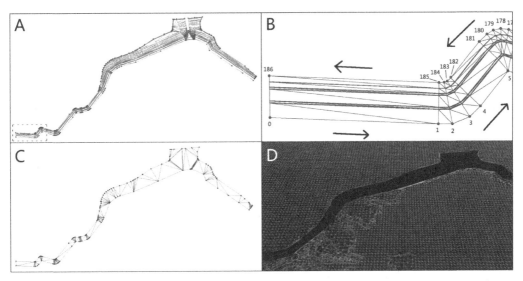

Fig. 6.6 Integrating the natural terrain with a 3D dam model. A panorama and local close-up of the blending boundary are in (A) and (B). The Delaunay triangulation within the boundary is in (C), and a constrained terrain triangulation for dam integration is in (D).

(2) Create a new point set of natural terrain elevations by first removing the vertices within and on the blending boundary, and then appending the above edge vertices. In this process, a Delaunay triangulation is first established for the simple concave polygon surrounded by the boundary (Fig. 6.6C). Then, the positional relationship between each terrain vertex and the boundary is determined via directed area calculations (Devadoss and O'Rourke, 2011) for the Delaunay mesh cells.

(3) Establish constrained Delaunay triangulation for the new point set and the blending boundary, according to the algorithm in Shewchuk and Brown

(Shewchuk and Brown, 2015). The resulting terrain model with a hole is used to seamlessly integrate the 3D dam model along the edge vertices (Fig. 6.6D).

6.2.4 Reservoir inundation extent analysis

Between the two types of inundation algorithms based on terrain models and water stage measurements, reservoir inundation extent analysis can be regarded as source flooding, since it generally reflects watershed sources that normally exceed the amount of precipitation in the reservoir area. The seeded region-growing algorithm and its deformations in limited memory-capacity or parallel computing environments are major solutions to source flooding (Shen et al., 2015; Zhang et al., 2014; Noman et al., 2001). These algorithms require three basic steps. First, preselect a processing area and a terrain mesh cell that is submerged as a seed point to start the computation. Second, recursively select a seed point and then searching for new submerged cells as seed points in all directions around this point, which are marked and stored in a stack. Each seed point in the top of the stack then undergoes another round of searching for new submerged cells after a popup operation, which removes the seed point from the stack. Finally, obtain final submerged cells when all seed points in the stack are removed. The seeded region-growing algorithms can accommodate various DEM data structures by changing search strategy, e.g., four directions around the seed point for a raster, and three directions for a triangulated irregular network (TIN), but rely on a pre-established neighboring relationship to find all adjacent terrain mesh cells, which is absent in 3D geometric models due to several difficulties, including intersected structural components and unexpected small defects. In addition, these algorithms assume that each terrain mesh cell can have only one value representing the elevation at that location, regarding the inundation estimation as a type of "binary" classification (i.e., 0 and 1). However, recent literature suggests advances in representing partially submerged terrain mesh cells, which can give a more detailed representation of inundation extent in riparian areas (Pan et al., 2013; Saksena and Merwade, 2015).

To address these limitations, this study enhances the seeded region-growing algorithm by delineating the inundation extent in partially submerged terrain mesh cells, and introducing the algorithm of intersection detection from computational geometry, so that submerged areas could spread from terrain mesh cells to associated hydraulic structures. The procedure is illustrated by the flow chart in Fig. 6.7 and detailed in the following sections.

Fig. 6.7 The flow chart for reservoir inundation extent estimation. The green, blue and yellow blocks correspond to the three main steps in Section 3.3.1 and 3.3.2, respectively.

6.2.4.1 Inundation on natural terrain

Instead of estimating a binary submerged result for each terrain mesh cell as in previous methods, this study takes a small step toward a more detailed representation of inundation extent. The details are as follows:

(1) Like the original seeded region-growing algorithm, a submerged terrain mesh cell in the center of the reservoir is preselected as a seed point.

(2) During the recursive process of region-growing from the three edges of a triangle and elevation comparison with the water stage, terrain cells are divided into three types: completely submerged, partially submerged, and not submerged (Fig. 6.8A). According to the number of triangle vertices that cross with the water stage, the partially submerged cells are further divided into four classes with additional two points affiliated to delineate the inundation extent (Fig. 6.8B). For classes a) and c), inundation points on the edge are obtained

Fig. 6.8 Inundation extent estimations on terrain triangular meshes. Three types of submerged results are shown in (A), and four classes of partially submerged cells are in (B).

through proportion calculation, whereas for b) and d), they are duplicated from triangle vertices.

(3) One or several closed curves are obtained by connecting the intersection points one by one, so that inundation extent on both the floodplains and islands are delineated.

6.2.4.2 Inundation on a 3D dam model

It is ideal to spread submerged areas from natural terrain to 3D geometric models to depict inundation extent on high parts of hydraulic structures. However, data structures of a 3D geometric model lack the neighboring relationship, making it difficult to find adjacent mesh cells around the seed point. The axis-aligned bounding box (AABB) is created for each dam triangular mesh cell, and the intersection detection of two bounding boxes can approximate the neighboring relationship of their host elements, which means if two AABBs intersect with each other, their host cells are regarded as adjacent, and vice versa (Fig. 6.9A). Therefore, all dam mesh cells below the water stage are first marked as submerged, and false cells, such as the structural components on the dorsal surface of embankment dams, are then filtered according to the AABB intersection detection, so that real inundation extent is delineated on the upstream face. The detailed procedures are as follows:

Fig. 6.9 Inundation extent estimations on a 3D dam model. A sketch map of establishing the neighboring relationship is in (A). The temporary result and final inundation extent are in (B) and (C), in which the blue cells are completely submerged, white cells are partially submerged, and the red line represents the water stage.

(1) Compare the water stage with all dam triangular mesh cells and insert those with at least one vertex below the stage into a temporary inundation collection (Fig. 6.9B). This leads to the problems that a) the elevation comparison cannot distinguish between the upstream face and the dorsal surface of the dam model, resulting in incorrect submerged results on both sides, and b) the internal structural components below the stage are also

submerged.

(2) Create AABBs for all the cells in the collection and perform the intersection detection by projecting the bounding boxes to the x-, y-, and z-axis, of which detailed algorithms are elaborated in Fabri and Pion (2009). If the projections of two AABBs overlap on the three axes, their host cells are marked as adjacent. Thus, the neighboring relationship of the temporary submerged mesh cells is prepared.

(3) Select a triangular mesh cell on the bottom of the upstream face as a seed point, and gradually spread the submerged area on the 3D dam model by searching the adjacent cells around the seed point. After filtering the false cells on the dorsal surface, the real inundation collection and submerged extent are obtained (Fig. 6.9C).

6.3 Results
6.3.1 An integrated reservoir terrain model

The above methodology is used to develop a terrain model of the Hengshan Reservoir based on three common tools, including an airborne LiDAR-derived DEM, bathymetric measurements, and Google imagery. The result embodies the topographic features of bathymetry, riparian areas, islands, and floodplains (Fig. 6.10A), but can hardly represent the morphology of associated hydraulic structures, such as the embankment dams, owing to the wide-coverage, but relatively coarse-resolution DEM data (Fig. 6.10B). According to previous studies, if the DEM points on the dam are deliberately elevated (Turner et al., 2013; Chen et al., 2012; Li et al., 2014), or transformed into constrained meshes to set up vertical walls (Shen et al., 2015; Zhang et al., 2014), they are water-

Fig. 6.10 The integrated reservoir terrain model. A panorama of the reservoir area, and local close-up of the embankment dams are in (A) and (B), while the same views, after integrating the 3D dam model, are in (C) and (D).

blocking. However, some specific properties, such as the thickness and length of the hypothetical walls, are unrealistic, complicating the delineation of the inundation extent on the upstream face of embankment dams.

Based on the natural terrain and 3D dam models, an integrated reservoir terrain model is established (Fig. 10C). The two models are assembled by extracting the blending boundary at the bottom of the dam, and replacing the terrain mesh cells below, so that a more realistic and coherent terrain surface can be prepared for inundation extent analysis (Fig. 10D), compared to the use of DEM and bathymetric measurements alone (Fig. 10B). The integrated model of the Hengshan Reservoir covers an area of 53.4 km^2, containing 85,500 elevation points, 168,400 mesh cells for the natural terrain, and 5,700 mesh cells for the embankment dams.

6.3.2 Reservoir inundation extent estimation

Taking rainfall and water stage measurements as input data, our previous studies have developed a flood stage forecasting model with the daily lead time for small reservoirs (Chen et al., 2017). Based on the model's trial operation in the Hengshan Reservoir, the inundation extent estimation in this study is first tested during an intense rainfall event when Typhoon Matmo hit southeastern China in 2014. When the rainfall began on July 25, the measured water stage was 33.1 m, and inundation covered the entire reservoir area (Fig. 6.11A) with a detailed representation of the water stage on the embankment dams (Fig. 6.11B). However, the flood stage forecasting model revealed that 20.3 million m^3 of the accumulated flood volume would enter by the next day, raising the stage to 37.2 m (Fig. 6.11C-D). To minimize flood risks to the dam structure and surrounding floodplains, appropriate water discharging schemes were

Fig. 6.11 Reservoir inundation extent estimation. The measured water stage is presented in (A) and (B), while the inundation extent prediction by the next day is in (C) and (D).

immediately performed to keep the stage below the flood limited water level (FLWL, 34.0 m).

To further analyze the computational efficiency of the proposed methods, a performance experiment was performed on the natural terrain model, the integrated terrain model, and the integrated terrain model with pre-processing on the neighboring relationship of dam mesh cells (Fig. 6.12). The experiment employed an ordinary PC with an Intel i7-4790 3.6GHz CPU, 16GB system memory, and a software environment of Python 2.7 in conjunction with Numpy packages in Windows 7. The computational time was only related to inundation extent estimations, excluding the data input, output and display.

The blue line in Fig. 6.12 indicates that estimations on the natural terrain model had the best performance from 24 to 30 m, but the model could not support higher water stages due to the DEM gaps on the embankment dams (see the asterisk). After integrating the 3D dam model, there was an average time increment of 244 ms below 27 m, and 1,352 ms from 28 to 30 m, which were 26% and 67% of the computational time for the natural terrain model. The two different time increments were due to the inclusion of the 3D dam model, especially the AABB intersection detection for the dam mesh cells. Further optimization by caching the intersection results of dam mesh cells into a square matrix of order n, where n is the number of dam mesh cells in the temporary inundation collection, and directly retrieving the adjacent cells during the inundation expansion, improved the efficiency by 27%, and reduced the time increment to 21% for water stages above 27 m. Therefore, the integrated terrain model has enabled inundation extent analysis at both low and high-water stages, and the performance is still acceptable, according to the de facto standard of "8-second rule" as articulated in modern software standards (Shewchuk et al., 2015).

Fig. 6.12 Performance analyses for inundation extent estimation. * Water flushes out from the DEM gaps, causing unreasonable inundation extent as the stage rises, if only the reservoir natural terrain model is used.

6.3.3 A comparison between the stage-storage curves

The stage-storage relationship, which serves as an indispensable tool for reservoir management, and mostly established from bathymetric measurements below the water surface and a polynomial extrapolation for the surrounding natural terrain and associated hydraulic structures (Mays, 2010), could also be obtained during inundation extent estimations in the integrated terrain model. Therefore, a new stage-storage curve for the Hengshan Reservoir was created and compared with the existing one in order to validate the effectiveness of the proposed methods for inundation extent estimation, especially at very high water stages.

After projecting the completely and partially submerged terrain mesh cells to the water surface, the volumetric extent of inundation is composed of generalized triangular prisms and their variants, so the reservoir water storage at a certain stage could be obtained by volumetric calculation and summation of these triangular prisms. For completely submerged cells, the water volume is the combination of a triangular prism and a pyramid (Fig. 6.13A). According to the spatial relationship between the water stage and three vertices, the partially submerged cells are divided into five classes for volumetric calculation (Fig. 6.13B). For classes a) and b), the volume is zero, whereas for classes c) and d), it is the tetrahedron on the vertex below, and two intersection points with

Fig. 6.13 Water volume calculation for the completely (A) and partially submerged terrain mesh cells (B)

the water surface. The volume of class e) is the combination of a triangular prism and a pyramid.

Fig. 6.14 compares the existing stage-storage curve with the new one, which is established based on the above volumetric calculations from 24 to 38 m with an interval of 0.1 m. From 24 to 30 m, the two curves approximately coincide with each other, but the new storage capacity is slightly less than the past. We speculate that this is due to the sediment in the past 10 years that has a slight impact on the bathymetric measurements at different times. The new curve is obviously longer than the existing one, but the difference between the extrapolated curve becomes larger as the water stage rises. The new stage-storage relationship appears to be more effective as it relies on coherent and realistic terrain surfaces, including the upstream face of embankment dams, rather than the assumption of a polynomial relationship in which the extrapolated trend is easily affected by the last few sample points (Cook and Stefanski, 1994). The integrated reservoir terrain model and proposed inundation methods are therefore better founded to determine the stage-storage relationship, and capable of providing reliable results for inundation extent analysis, especially at very high water stages.

Fig. 6.14 The comparison between the stage-storage curves. The red and green dotted lines are respectively created from bathymetric measurements and a polynomial extrapolation, and the blue line is established based on the integrated reservoir terrain model and proposed inundation methods.

6.4 Discussion

Highly resolved terrain modeling plays a decisive role in inundation simulations and flood risk assessments, especially for studies at meso- and micro-scales. This study refrains from using resource-intensive techniques, such as terrestrial and airborne topo-bathymetric LiDAR (Sampson et al., 2012; Andersen et al., 2017), but employs a wide-coverage, but relatively coarse-resolution DEM, bathymetric measurements, and a 3D dam model from easily available data sources to create terrain models for small reservoirs. Compared with previous techniques that rely heavily on field surveys to set up the buffer polygons and interpolation weights in riparian areas (Cook and Merwade, 2009; Conner and Tonina, 2014; Merwade et al., 2008), the proposed method utilizes optical and topographic information to divide the reservoir area into several subregions, so that only ambiguous areas require these additional procedures (Fig. 4). Although photogrammetric mapping methods, such as feature extraction, are prevalent and robust to identify land-water boundaries (Gruen et al., 2012), this study employs Google imagery during the intermediate steps, owing to several advantages with regard to small reservoirs: a) the spatial resolution is higher than most of the freely downloadable datasets, e.g., Landsat TM Imagery (30 m resolution); and b) instead of pre-screening for ideal conditions, such as clear skies and low water vapor, the image quality is ensured by Google Earth. The natural terrain is integrated with a 3D dam model to eliminate the simplifications and distortions related to the morphology of associated hydraulic structures. Compared with building blocks or hypothetical walls (Shen et al., 2015; Turner et al., 2013; Zhang et al., 2014; Li et al., 2014), the use of a 3D geometric model avoids extra settings related to the dam appearance, and is more realistic for inundation extent analysis.

Although AABB intersection detection has enabled source flooding to cover associated hydraulic structures, there are two potential problems. First, the computational time, especially for water stages higher than 27 m, is significant (67% more than the baseline). Even if the neighboring relationship is cached in a matrix to avoid real-time intersection detection, it still increases by 21% on average (Fig. 6.12). Therefore, to enhance the proposed inundation methods in practical applications, further improvement in computational efficiency is recommended, especially when involving large-scale, high-resolution DEMs, and sophisticated 3D geometric models. Some referential examples are spatial indexing, such as R- or R*-trees, and the corresponding traverse strategy based on the Hilbert curve, which are proved to accelerate nearest neighbor searches in 3D geometric objects (Sharifzadeh and Shahabi, 2010; Chen and Chang, 2011; Chang et al., 2016).

Second, it is still difficult to distinguish the surface and internal structures of the 3D dam model, resulting in unreasonable inundation expansion from the upstream face to the internal components (Fig. 6.3B and 6.9C). An appropriate solution may be to employ an enhanced 3D geometric model, which has incorporated semantic information of the structural components, such as topological relationships and physical properties. Accordingly, current advances in building information modeling (BIM) have encouraged a computable representation of all the physical and functional characteristics of a building, which is intended to bridge the information gap between project design, construction, and operation (Amirebrahimi et al., 2016; Amirebrahimi et al., 2016). Meanwhile, in contrast to purely geometrical/graphical models, such as VRML, X3D, and OBJ, the Geography Markup Language 3 (GML3) of the Open Geospatial Consortium (OGC) and its application schema, namely CityGML, have considered both structural and semantic aspects of a 3D geometric model, enabling advanced analysis and visualization in various application domains associated with urban environments (Biljecki et al., 2015; Gröger and Plümer, 2012). Based on these interoperable 3D models, more realistic inundation extent estimations and associated risk assessments for hydraulic structures are expected in future research.

Comparing flood routing simulations based on hydrologic and hydrodynamic models, inundation extent analysis using a horizontal water surface has the advantages of little required input data and fast calculation speed, and has been employed by several rapid flood simulation models (Zhang and Pan, 2014; CH2M HILL, 2017; Krupka, 2009). Similarly, the horizontal water surface in this study is first obtained from a pre-established flood stage forecasting model (Chen et al., 2015), and then used to estimate inundation situations in the Hengshan Reservoir. Although the generalized result only simulates the final submerged extent, and ignores the hydraulic gradients in flowing floods, it is practical to reservoir flood control management. In addition, the proposed enhancements to the source flooding algorithm, including setting up closed curves in partially submerged cells, and spreading submerged areas to a 3D geometric model, are equally applicable to flood routing simulations based on on-site observations or 2D/3D hydrodynamic models. This is because the difference between a horizontal and gradient water surface only lies in obtaining the water stage for each terrain mesh cell to determine whether that cell is submerged or not (Fig. 6.8). Therefore, the proposed methods can also inform other flood inundation scenarios such as river channels and urban areas.

This study employs several commercial GIS tools and open-source software packages during the terrain integration and inundation extent analysis. For example, ArcGIS is used for integrating the bathymetry with surrounding topography, and delineating the blending boundary on the 3D dam model, Numpy packages help to establish Delaunay triangulations from discrete elevation point sets, and to implement the enhanced source flooding algorithm, and 3DS Max serves to convert the 3D dam model into an open-format surface triangulation, and to create AABBs for temporary submerged mesh cells. However, these routines are not necessarily dependent on the above tools and packages, since their principles have been implemented by many other off-the-shelf software and libraries, making the proposed methods feasible in various operating systems and software environments. Nevertheless, there is no single solution, e.g., ArcGIS Plugin, to facilitate inundation extent analysis in similar reservoirs, especially when involving the embankment dams, which suggests the need to assemble the above routines into integrated and user-friendly interfaces, to improve the resilience of small reservoirs with respect to floods and their adverse impacts.

6.5 Conclusion

Inundation extent analysis is a crucial exercise for flood risk mitigation, although it is less developed for applications to small reservoirs. This study focuses on the combined use of various tools and methods to achieve rapid and effective estimation of inundation extent for both small reservoirs and their associated hydraulic structures. The reservoir terrain surface can be more realistic and coherent, if topography, bathymetry and a 3D dam model are integrated. The source flooding algorithm can present a detailed water stage on the embankment dams, if the AABB intersection detection from computational geometry is introduced to spread submerged areas from terrain mesh cells to a 3D geometric model. These key findings are tested by a historical rainfall event and a performance experiment in the Hengshan Reservoir in Southeastern China. They are also proved to support extreme water stages, and better-founded to determine the reservoir stage-storage relationship. To eliminate false inundation results on the internal structures of a 3D dam model, and enhance the proposed methods in practical applications of similar small reservoirs, further research should include incorporation of enhanced 3D geometric models with semantic information, establishing spatial indexing to improve computational efficiency, and assembling the routines into comprehensive, synthetic and user-friendly interfaces.

References

Alcrudo, F., 2004. Mathematical modelling techniques for flood propagation in urban areas, Project report: IMPACT Project.

Amirebrahimi, S. A., Rajabifard, P., Mendis, T. N., 2016. "A framework for a microscale flood damage assessment and visualization for a building using BIM-GIS integration." *Int. J. Digit. Earth* 9(4): 363-386.

Amirebrahimi, S., Rajabifard, A., Sabri, S., et al., 2016. "Spatial information in support of 3D flood damage assessment of buildings at micro level: A review, ISPRS Annals of Photogrammetry." *Remote Sensing and Spatial*

Information Sciences 4(2): 73-81.

Andersen, M.S., Gergely, Á., Al-Hamdani, Z., 2017. "Processing and performance of topobathymetric lidar data for geomorphometric and morphological classification in a high-energy tidal environment." *Hydrol. Earth Syst. Sc.* 21(1): 43-63.

Biljecki, F., Stoter, J., Ledoux, H., 2015. "Applications of 3D city models: State of the art review." *ISPRS Int. J. of Geo-Inf.* 4(4): 2842-2889.

Chang, Y. H., Liu,Y. T., Tan,Y.Y.,2016. "Landmark-based summarized messages for flood warning." *Transactions in GIS* 21(5): 847-861.

Chen, A.S., Evans, B., Djordjević, S., et al., 2012. "A coarse-grid approach to representing building blockage effects in 2D urban flood modelling." *J. Hydrol.* 426-427: 1-16.

Chen, H.L., and Chang, Y. I., 2011. "All-nearest-neighbors finding based on the Hilbert curve." *Expert Syst. Appl.* 38(6): 7462-7475.

Chen, J., Wang, D., Zhao, R., et al., 2014. "Fast updating national geo-spatial databases with high resolution imagery: China's methodology and experience." *Int. Arch. Photogramm. Remote Sens. Spatial Inf. Sci.*XL-4: 41-50.

Chen, W., He, B., Nover, D., et al., 2017. "Spatiotemporal patterns and source attribution of nitrogen pollution in a typical headwater agricultural watershed in Southeastern China." *Environ. Sci. Pollut. Res.* 25: 2756-2773.

Chen, W., He, J., Ma, C., et al., 2017. "WebGIS-based flood control management system for small reservoirs: A case study in the lower reaches of the Yangtze River." *J. Hydroinform.*19(2): 299-314.

Conner, J.T., and Tonina, D., 2014. "Effect of cross-section interpolated bathymetry on 2D hydrodynamic model results in a large river." *Earth Surf. Proc. Land.* 39(4): 463-475.

Cook, A., and Merwade, V., 2009. "Effect of topographic data, geometric configuration and modeling approach on flood inundation mapping." *J. Hydrol.* 377(1/2): 131-142.

Cook, J.R., and Stefanski, L.A., 1994. "Simulation-extrapolation estimation in parametric measurement error models." *J. Am Stat. Assoc.* 89(428): 1314-1328.

Curtarelli, M., Leão, I. J., Ogashawara, J., et al., 2015. "Assessment of spatial interpolation methods to map the bathymetry of an Amazonian hydroelectric reservoir to aid in decision making for water management." *ISPRS Int. J. Geo-inf.* 4(1): 220-235.

Demir, I., and Krajewski, W. F., 2013. "Towards an integrated flood information system: centralized data access, analysis, and visualization." *Environ. Modell. Softw.*50: 77-84.

Devadoss, S.L., and O'Rourke, J., 2011. *Discrete and computational geometry.* New Jersey: Princeton University Press.

Döll, P., Jiménez-Cisneros, B., Oki, T., et al., 2015. "Integrating risks of climate change into water management." *Hydrolog. Sci. J.* 60(1): 4-13.

Fabri, A., and Pion, S., 2009. "CGAL: The computational geometry algorithms library." In *17th ACM SIGSPATIAL international conference on advances in geographic information systems,* Seattle, Washington, November 4-6.

Giorgio, G.A., Ragosta, M., Telesca, V., 2017. "Climate variability and industrial-suburban heat environment in a Mediterranean area." *Sustainability* 9(5): 775.

Gröger, G., and Plümer, L., 2012. "CityGML-Interoperable semantic 3D city models." *ISPRS J. Photogramm.* 71: 12-33.

Gruen, A., Baltsavias, E., Henricsson, O., 2012. *Automatic extraction of man-made objects from aerial and space images (II).* Switzerland: Birkhauser.

Hirabayashi, Y., Mahendran, R. Koirala, S., et al., 2013. "Global flood risk under climate change." *Nat. Clim. Change* 3(9): 816-821.

Hoover, C., and methodology. A., 2006. "A Methodology for determining response time baselines: Defining the "8 Second" Rule." In *Proceedings of the annual international conference of the computer measurement group (CMG2006),* Reno, Nevada.

Janardan, R., Smid, M., Dutta, D., 2005. "Geometric and algorithmic aspects of computer-aided design and manufacturing." In DIMACS workshop computer aided design and manufacturing. Piscataway, NJ, October 7-9.

Krupka, M., 2009. "A rapid inundation flood cell model for flood risk analysis." Doctoral thesis, Heriot-Watt University.

Lai, J.S., Chang,W.Y., Chan, Y.C., et al., 2011. "Development of a 3D virtual environment for improving public participation: Case study-The Yuansantze flood diversion works project." *Adv. Eng. Inform.* 25(2): 208-223.

Lay-Ekuakille, A., Telesca, V., Ragosta, M., et al., 2017. "Supervised and characterized smart monitoring network for sensing environmental quantities." *IEEE Sens. J.*17(23): 7812-7819.

Li, Z., Wu, L., Zhu, W., et al., 2014. "A new method for urban storm flood inundation simulation with fine CD-TIN surface." *Water* 6(5): 1151-1171.

Liu, X.J., Zhong, D.H., Tong, D.W., et al., 2016. "Dynamic visualisation of storm surge flood routing based on three-dimensional numerical simulation." *J. Flood Risk Manag.*11(1): 729-749

Manian, D., Kaihatu, J. M., Zechman, E. M., 2012. "Using genetic algorithms to optimize bathymetric sampling for predictive model input." *J. Atmos. Ocean. Tech.*29(3): 464-477.

Mays, L.W., 2010. *Water resources engineering.* New Jersey: John Wiley & Sons.

Merwade, V., Cook, A., Coonrod, J. G., 2008. "GIS techniques for creating river terrain models for hydrodynamic modeling and flood inundation mapping." *Environ. Modell. Softw.* 23(10): 1300-1311.

Neubert, M., Naumann, T., Hennersdorf, J. et al., 2014. "The geographic information system-based flood damage simulation model HOWAD." *J. Flood Risk Manag.* 9(1): 36-49.

Noman, N.S., Nelson, E.J., Zundel, A.K., 2001. "Review of automated floodplain delineation from digital terrain models." *J. Water Res. Pl-ASCE.*127(6): 394-402.

Pan F., Liao, J., Li, X., et al., 2013. "Application of the inundation area-lake level rating curves constructed from the SRTM DEM to retrieving lake levels from satellite measured inundation areas." *Comput Geosci.* 52: 168-176.

Pisaniello, J.D., Dam, T.T., Tingey-Holyoak, J.L., 2015. "International small dam safety assurance policy benchmarks to avoid dam failure flood disasters in developing countries." *J. Hydrol.* 531: 1141-1153.

Rosenberg, D.M., McCully, P., Pringle, C.M., 2000. "Global-scale environmental effects of hydrological alterations: introduction." *BioScience* 50(9): 746-751.

Saksena, S., and Merwade, V., 2015. "Incorporating the effect of DEM resolution and accuracy for improved flood inundation mapping." *J. Hydrol.* 530: 180-194.

Sampson, C.C., Fewtrell, T.J., Duncan, A., et al., 2012. "Use of terrestrial laser scanning data to drive decimetric resolution urban inundation models." *Adv. Water Resour.* 41: 1-17.

Sanders, B.F., Schubert, J.E., Detwiler, R.L. et al., 2010. "ParBreZo: A parallel, unstructured grid, Godunov-type, shallow-water code for high-resolution flood inundation modeling at the regional scale." *Adv. Water Resour.* 33(12): 1456-1467.

Sarhadi, A., Soltani, S., Modarres, R., 2012. "Probabilistic flood inundation mapping of ungauged rivers: Linking GIS techniques and frequency analysis." *J. Hydrol.* 458: 68-86.

Selvanathan, S., and Dymond, R.L., 2010. "FloodwayGIS: An ArcGIS visualization environment to remodel a floodway." *Transactions in GIS* 14(5): 671-688.

Sharifzadeh, M., and Shahabi, C., 2010. "Vor-tree: R-trees with voronoi diagrams for efficient processing of spatial nearest neighbor queries." *Proceedings of the VLDB Endowment* 3(1-2):1231-1242.

Shen, D., Rui, Y., Wang, J., et al., 2015. "Flood inundation extent mapping based on block compressed tracing." *Comput Geosci.* 80: 74-83.

Shewchuk, J.R., and Brown, B.C., 2015. "Fast segment insertion and incremental construction of constrained Delaunay triangulations." *Comp. Geom-Theor. Appl.* 48(8): 554-574.

Snyder, G.I., 2012. National enhanced elevation assessment at a glance (No. 2012-3088), US Geological Survey.

Song, W.Z., Jiang, Y.Z., Lei, X.H., et al., 2015. "Annual runoff and flood regime trend analysis and the relation with reservoirs in the Sanchahe River Basin, China." *Quatern. Int.* 380-381: 197-206.

Teng, J., Jakeman, A.J., Vaze, J., et al., 2017. "Flood inundation modelling:

A review of methods, recent advances and uncertainty analysis. Environ." *Modell. Softw.* 90: 201-216.

Turner, A.B., Colby, J.D., Csontos, R.M., et al., 2013. "Flood modeling using a synthesis of multi-platform LiDAR data." *Water* 5(4): 1533-1560.

Vacondio, R., Dal, A., Mignosa, P., 2014. "GPU-enhanced finite volume shallow water solver for fast flood simulations." *Environ. Modell. Softw.* 57: 60-75.

Walt, S.V.D., Colbert, S.C., Varoquaux, G., 2011. "The NumPy array: a structure for efficient numerical computation." *Comput. Sci. Eng.* 13(2): 22-30.

World Commission on Dams, 2000. *Dams and development: A new framework for decision-making.* London: Earthscan Publications.

Yadav, B., Ch, S., Mathur, S., et al., 2016. "Discharge forecasting using an online sequential extreme learning machine (OS-ELM) model: A case study in Neckar River, Germany." *Measurement* 92: 433-445.

Yamazaki, D., Baugh, C.A., Bates, P.D., et al., 2012. "Adjustment of a spaceborne DEM for use in floodplain hydrodynamic modeling." *J. Hydrol.* 436: 81-91.

Zhang, S., and Pan, B., 2014. "An urban storm-inundation simulation method based on GIS." *J. Hydrol.* 517: 260-268.

Zhang, S., Wang, T., Zhao, B., 2014. "Calculation and visualization of flood inundation based on a topographic triangle network." *J. Hydrol.* 509: 406-415.

Zhang, S., Xia, Z., Wang, T., 2013. "A real-time interactive simulation framework for watershed decision making using numerical models and virtual environment." *J. Hydrol.* 493: 95-104.

Chapter
7

Developing a WebGIS-based Flood Control Management System for Small Reservoirs

7.1 Introduction

To modify the uneven distribution of water resources in both time and space, more than 847,000 reservoirs have been constructed globally in the last 100 years, and approximately 95% of these are small reservoirs (Song et al., 2015). They not only provide water, hydroelectric energy and irrigation but also stabilize extreme inflows to mitigate floods or droughts. Due to the natural uncertainties, public concern about the small reservoirs and the specific efforts for flood control management has increased in recent years (Chang and Chang, 2006; Rodrigues et al., 2012).

Reservoir flood control management is a historic issue regarding the use of structural and non-structural measures to optimize flood discharge and reduce flood risks. It requires simultaneous considerations about the hydrologic, geotechnical, environmental and behavioural aspects. Among them, the flood forecast, including volume, peaks and duration, is one of the key factors. Many flood-prone countries have employed forecasting systems since the 1990s. There includes the National Weather Service River Forecasting System (NWSRFS) for 13 main rivers across the U.S. (Burnash and Singh, 1995), the Delft Flood Early Warning System (Delft-FEWS) for several European countries (Werner et al. 2009), the integrated flood control management system for 69 large reservoirs of China (Cheng and Chau, 2004), etc. These early examples provide flood control decision support by establishing chart-based user interfaces around the hydrological and hydraulic models used. But their application and usage require a good understanding of flood processes, which is feasible for national operational centres (Werner et al., 2013; Cools et al., 2016), but difficult for small reservoir managers with uneven levels of hydrological knowledge.

Recent development in information technology has provided substantial opportunities to enhance the early warning and flood control at different spatial scales. There are global systems for upcoming floods in large world river basins (Alfieri et al., 2013; Wu et al., 2014), and continental systems that complement the national systems with medium- and long-range forecasts for transboundary rivers and lakes (Roo et al., 2011; Werner et al., 2013; Thiemig et al., 2015). Flood control systems for local areas have widely incorporated web-based technologies. Demir and Krajewski (2013) provided an integrated online platform with flood forecasts and inundation maps. Horita et al. (2015) combined the monitoring data with volunteered geographic information for river flood risk management. The above systems combined with desktop- and web-based GIS are proved to be integrative and cooperative environments for river basins and urban watersheds, so that researchers, decision-makers and the general public are better involved. But practical flood control management usually requires targeted and deterministic data to help formulate discharging schemes, especially for the numerous and widely distributed small reservoirs. In order to keep the results concise and effective for reservoir managers, the system has to cover multidisciplinary knowledge within, but presents as user-friendly interfaces.

Many rainfall-runoff models have been developed to generalize hydrologic processes and offer flood forecasts. Some are lumped conceptual models, such as the Sacramento model, the Xinanjiang model, and the tank model (Zhao, 1992; Lü et al., 2013). They are applied in many regions but typically have more than 10 parameters that require gradual adjustment to the measured streamflow data (Chang and Chang, 2006). Particularly, the Soil Conservation Service Curve Number (SCS-CN) model is designed for small watersheds (Mishra and Singh, 2003). Due to the simple structure and clearly stated assumptions, relatively small amounts of data are required when using this model. However, the spatial analysis for watershed characteristics, such as elevation, soil type, and land use, is essential to obtain more accurate CN values, rather than depending on the tables developed for U.S. conditions.

To represent the spatial conditions of underlying surface and meteorological data, GIS has enabled the models to have physically distributed characteristics. The LISFLOOD model is an example that simulates the spatial and temporal pattern of river discharge in large basins. The TOPKAPI model is applied to medium-sized river basins, which transforms rainfall-runoff routing processes into non-linear reservoir differential equations. The HEC-HMS model supports numerous infiltration loss parameterizations to calculate overland flow runoff (Vieux, 2001; Ciarapica and Todini, 2002; Van Der Knijff et al., 2010). The selection and application of rainfall-runoff models is one of the most important tasks for reservoir flood control management, since the specific flood forecast relies not only on the practical hydrological measurements but also the available time-range to guide flood discharge.

The Yangtze River has been subjected to flooding throughout history, especially for the provinces and municipalities at its lower reaches, which are of great economic importance for China (Cheng and Chau, 2004). Due to the intensive rainfall events and humid climates during the flood season, inflow floods rapidly raise the water levels and cause dam failure risks (Zhao et al., 2014). Although flood control management is available for large reservoirs, it is inadequate for small reservoirs. Instead of using advanced forecasting and decision-support tools, these reservoirs primarily rely on the long-term field experience of the operation personnel who prefer to maintain a higher capacity due to economic benefits, but this approach introduces a significant number of uncertainties regarding the dam safety. Meanwhile, the hydrological measurements are mostly limited to the water level logged by days or hours, and rainfall in or near the catchment, which limits the direct usage of existing rainfall-runoff models.

To improve this situation, a WebGIS-based flood control management system has been developed since 2014, and the Hengshan Reservoir in Jiangsu Province is selected as the study area for the typical hydrologic and reservoir project conditions in the lower reaches of the Yangtze River. To date, it has completed a trial operation and will be gradually applied to other small reservoirs in the vicinity.

In order to provide useful reference to the flood control and water resource utilization in humid regions, this study focuses on the design and implementation of the WebGIS-based system. It tries to use multidisciplinary knowledge to establish user-friendly tools to bridge the gap between large-scale flood control systems and practical management of small reservoirs. It also investigates the capability to present concise flood forecasts based on limited hydrological measurements to guide the flood discharge during intensive rainfall in the flood season. Typical rainfall events in the Hengshan Reservoir are chosen to estimate the model parameter values, and approximately one hundred raining days are used to analyse the forecast accuracy. After describing a real flood discharging example, future improvements and recommendations for the system are discussed.

7.2 Methodology
7.2.1 Study area and materials

The catchment of the Hengshan Reservoir covers an area of 154.8 km^2 with an average elevation of 300 m, which is obtained by ArcGIS Hydrology toolset and the 30-m-resolution Aster DEM. There is a hydrological station on the dam and five rain gauges evenly distributed in the catchment (Fig. 7.1). The automatic gauging system records the hourly rainfall and water level. According to the measurements, the flood peak inflows approximately 8 hours after the intensive rainfall. The reservoir flood-limited water level (FLWL) is 34.0 m, the normal water level (NWL) is 35.0 m, and the maximum flood level is 37.0 m. Particularly, the FLWL is the most significant indicator to offer adequate storage for flood prevention during the flood season, and the maximum discharge is about 550 m^3/s, which takes 4 hours to lower the stage from the NWL to the FLWL.

Fig. 7.1 Study area of the Hengshan Reservoir in the lower reaches of the Yangtze River

There are two classes of basic materials used in this study. One is necessary for flood level forecasts, as shown in Table 7.1, and another is optional to enrich the map browse and water level monitoring, which contains an online geographic base map, 30-m DEM, dam construction designs, photos taken at key locations, and video surveillance.

Table 7.1 Basic materials for small reservoir flood level forecasts

Data type	Content and resolution	Source
Rainfall measurements	Hourly, from 2006 to 2015	Automatic gauging system
Water level measurements	Hourly, from 2006 to 2015	Automatic gauging system
Reservoir basic information	Stage-storage curve and average daily water consumption	Reservoir administrative division
Evaporation data	Daily, from 2006 to 2012	China National Meteorological Information Centre

7.2.2 Adopting agile methods to involve different participants

Providing accurate and efficient inflow flood forecasts during intensive rainfall to support flood discharge is the main purpose of the system. Meanwhile, some characteristics of small reservoirs have to be considered, such as limited hydrological measurements, fast rainfall-runoff processes and uncertainties of geospatial data composition. Therefore, the system design and implementation require a combination of multidisciplinary knowledge from different fields, such as reservoir management, flood discharge, hydrological modelling, map development and software integration.

Agile software development methods are currently adopted in many cross-regional and cross-domain practices to cope with crises, such as rising complexity and extended cycles (Brhel et al., 2015). Compared with traditional planning-based methods, it values a progressive and iterative approach to achieve effective results. Following the principles of agile methods, this study uses scenario description, visual design and rapid prototyping to keep the system available during the entire process of design, implementation and application. Thus, different participants are initiatively involved, and make suggestions to improve the system based on their respective expertise.

Specifically, two process assistant tools are used from the initial design phases. One is an online collaborative tool named Mingle to help resolve the overall objective into several detailed sub-tasks in each iterative development cycle (Fig. 7.2A). Another is a prototyping tool named Mockups to help visualize the abstract concepts and embody their interactions in webpage wireframes (Fig. 7.2B), so that reservoir managers and hydrological experts could work in a what-you-see-is-what-you-get (WYSIWYG) environment.

Fig. 7.2 Adopting agile methods during the entire development process. Experts from different fields fully participate to resolve the objectives based on function requirement cards (A), while the corresponding prototypes are used to keep the designs available and shareable (B).

7.2.3 System architecture design

The system consists of three parts: the data tier, the service tier and the decision-support tier (Fig. 7.3). It is integrated using .Net Framework in VS2012 and adopts a loosely coupled development approach of different tools and languages to enable collaborative implementation by the experts from different fields. In the data tier, several types of basic data are collected from stable and mobile devices, which include rainfall and water level measurements, as well as video streaming and multimedia files, such as dam construction designs and photos. These data are respectively stored in rational tables of Oracle 11 and the file system of Windows Server 2008. In addition, external map tiles are used by invoking static Google Maps APIs. In the service tier, the flood level forecasting model and its integration with flood discharging schemes are developed using the FORTRAN language. The records over the past days, as well as the current hourly data, are provided by the water level and rainfall data service. In addition, a similar rainfall process query retrieves the 5 consecutive days in history whose cumulative rainfall is approximate to the current amount,

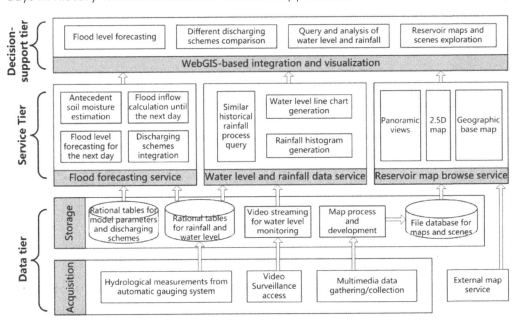

Fig. 7.3 Architecture of the system

so that the reservoir manager has a real discharging instance for reference in the flood season. The decision-support tier is in a WebGIS-based application integrated by Ext JS, JQuery and TeeChart. This tier provides a routine overview of the reservoir project, as well as water level visualization and flood discharging schemes comparison.

7.2.4 Development of the flood level forecasting model

7.2.4.1 Overall structure and procedure

The forecasting model aims to evaluate the accumulated inflow flood volume and flood level in the future based on occurred rainfall so that different discharging schemes can be compared to optimize the final capacity. The lead time has to cover inflow flood peak to ensure the validity of the results. Thus, a daily forecast range is used to adapt practical flood control management and other similar small reservoirs that possess only daily hydrological measurements. Following the schematic diagram of the Xinanjiang model (Zhao, 1992; Lü et al., 2013), it is established by generalizing the rainfall-runoff process into four parts. They are rainfall and evaporation, a double-layered soil model, surface and subsurface runoff, as well as flood discharge and water consumption (Fig. 7.4A). The detailed concepts are as follows:

Fig. 7.4 Overall structure and procedure of the flood level forecasting model. The rainfall-runoff generalization and four primary parts are shown in (A), and the five main calculation steps are shown in (B).

(1) The Thiessen polygons of the rainfall gauges are established to obtain total rainfall amounts in the catchment.

(2) The double-layered soil moisture model is used to estimate the rainfall-runoff. The upper layer receives rainfall and quickly gathers surface runoff, while the lower layer receives infiltration from the above layer and gradually forms subsurface runoff.

(3) The evaporation assessment depends on the rainy conditions. On a rainy day, it is ignored due to the humid conditions. On a non-rainy day, it assumes that there is a constant consumption in the upper soil layer.

(4) Flood inflow contains the surface and subsurface runoff. The reservoir water consumption, such as drinking, irrigation and leakage, is generalized as flood discharge. So, there is a minimum daily discharge amount.

The model contains eight parameters, which are shown in the first three columns of Table 7.2, and the overall forecasting procedure is presented in Fig. 7.4B with detailed descriptions as follows:

Table 7.2 Model parameters and their values for the Hengshan Reservoir

Parameter	Description	Unit	Value
W_m	Maximum antecedent soil moisture	mm	90.4
W_{Um}	Maximum antecedent moisture of the upper soil layer	mm	31.8
W_{Lm}	Maximum antecedent moisture of the lower soil layer	mm	58.6
N	Number of days before for soil moisture estimation		14
R_b	Daily subsurface runoff volume	mm	5.3
E	Evaporation consumption on a non-rainy day	mm	4.2
A	Catchment area	m^2	1.548×10^8
OF_{min}	Minimum daily discharge amount	m^3	2.5×10^5

(1) Retrieve the evaporation constant and daily rainfall amounts during N days before the forecasting day.

(2) Estimate the soil moisture of the two layers from N day before to one day before.

(3) Calculate the accumulated inflow flood volume based on the surface and subsurface runoff.

$$IF = R_1 \times A \times 10^{-3} \qquad (1)$$

where IF is the accumulated inflow flood volume (m^3) and R_1 is the total runoff volume (mm).

(4) If the forecasted flood level based on the reservoir stage-storage curve is higher than the FLWL during the flood season, flood discharge is required

immediately. The total outflow volume can be calculated as follows:

$$OF = OF_{min} + \sum_{i=1}^{n} v_i t_i \qquad (2)$$

where of is the total outflow volume (m^3), v_i is the flood discharge during a particular period (m^3/s), and t_i is the corresponding duration (s).

(5) Finally, the discharged water level is obtained by the actual change of storage capacity.

7.2.4.2 Antecedent soil moisture estimation

Based on the saturation-excess runoff theories in humid regions (Vieux, 2001), a double-layered soil moisture model is designed for the rainfall-runoff estimation. It uses a concept model of two boxes, with one placed inside the other (Fig. 7.5). The small box directly receives rainfall and evaporates on a non-rainy day. If it is filled, rainfall overflows into the big box. If both of them are filled, then extra rainfall spills over. Thus, the soil moisture estimation of the two layers is presented below:

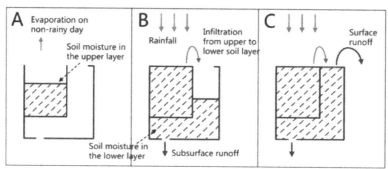

Fig. 7.5 Sketch map of the antecedent soil moisture estimation

(1) On a non-rainy day,

$$W_{U2} = W_{U1} - E, \; W_{L2} = W_{L1} - R_b \qquad (3)$$

where W_{U2} and W_{L2} are the soil moisture values (mm) of the two layers, while W_{U1} and W_{L1} are the same parameters one day before.

(2) On a rainy day, the evaporation is ignored; thus,

$$\text{If } P + W_{U1} \leqslant W_{Um}, \text{ then } W_{U2} = W_{U1} + P, \; W_{L2} = W_{L1} - R_b$$

$$\text{If } P + W_{U1} > W_{Um}, \text{ then } W_{U2} = W_{Um}, \; W_{L2} = W_{L1} + (W_{U1} + P - W_{Um}) - R_b \;(4)$$

where P is the rainfall amount (mm).

7.2.4.3 Rainfall-runoff calculation

Analogous to a series of tanks in a vertical array to express water storage, infiltration and runoff in the tank model, the rainfall-runoff calculation uses several laterally connected soil moisture models to express the hydrological process (Fig. 7.6). The initial soil moisture is set to zero on the N-th day before to start the calculation process. During the N days, surface and subsurface runoff gradually inflows and presents as water level measurements, while the remaining soil moisture is used to calculate the runoff volume on the forecasting day. The detailed procedures are as follows:

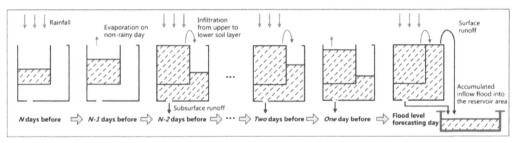

Fig. 7.6 Sketch map of the rainfall-runoff calculation

First, beginning on the-th day before, the following can be expressed:

$$\text{If } 0 \leqslant P_N \leqslant W_{Um}, \text{ then } W_{UN} = P_N, \quad W_{LN} = 0$$
$$\text{If } W_{Um} < P_N \leqslant W_m, \text{ then } W_{UN} = W_{Um}, \quad W_{LN} = P_N - W_{UN}$$
$$\text{If } W_m < P_N, \text{ then } W_{UN} = W_{Um}, \quad W_{LN} = W_{Lm} \tag{5}$$

Second, the soil moisture is estimated day by day until the forecasting day is reached.

$$\text{If } W_{Ui} + P_i \leqslant W_{Um}, \text{then } W_{Ui} = W_{U(i+1)} + P_i, \quad W_{Li} = W_{L(i+1)} - R_b$$
$$\text{If } W_{Ui} + P_i > W_{Um}, \text{then } W_{Ui} = W_{Um}, \quad W_{Li} = W_{U(i+1)} + P_i - W_{Ui} + W_{L(i+1)} - R_b$$
$$\text{If } W_{Li} > W_{Lm}, \text{ then } W_{Li} = W_{Lm} \tag{6}$$

Finally, the result flood inflow is composed of the surface and subsurface runoff.

$$\text{If } P + W_0 \leqslant W_m \text{ then } R_d = 0$$
$$\text{If } P + W_0 > W_m \text{ then } R_d = P + W_0 - W_m$$
$$R_l = R_d + R_b \tag{7}$$

where R_d is the surface runoff volume (mm).

7.2.5 Development of reservoir maps

Due to the large amounts and remote distribution of small reservoirs, fast location identification for the reservoir area and the spatial layout of important hydraulic structures are useful for a flood control management system. Therefore, three types of online maps are integrated, so that reservoir managers can easily acquire the background information without geospatial expertise. There includes a geographic base map, a 2.5D map and panoramic views. These maps actually form a three-angle perspective of vertical downward, obliquely downward and horizontal directions for the entire reservoir area. The development method avoids using a bundled WebGIS software platform, such as ArcServer; thus, the map browsing functions are more flexible according to the availability of spatial materials of a particular reservoir. A technical roadmap is shown in Fig. 7.7, and the details are as follows:

Fig. 7.7 Technical roadmap of the reservoir map development

(1) The 3D terrain model is generated by overlaying the image of the reservoir area onto the DEM data, while the 3D dam model is designed and textured from dam construction CAD drawings in 3D Studio Max. The matching and blending of their relative positions establish the reservoir 3D scene. After that, the 2.5D map tiles are acquired by rendering at a 45-degree angle downward via the 3D Studio Max slicing tool.

(2) Based on the photos taken in six directions of forward, backward, left, right, above and below at key locations in the reservoir area, local panorama photos are stitched and generated. After setting their positions in the 3D scene, the panorama views are created using Pano2VR and exported to a single flash file to embed into web pages.

(3) The geographic base map is a direct reference of external map services

from Google Maps. It is used as the very basic material to explore the reservoir area when the above multimedia files are unavailable.

7.3 Results and discussion
7.3.1 Model parameter acquisition and verification

According to the theories of the SCS-CN model, the antecedent soil moisture can be approximated as zero if intensive rainfall occurs over the entire watershed after several days of drought (Shi et al., 2009). Thus, the value of the Hengshan Reservoir catchment is analysed by choosing typical rainfall events based on three empirical principles: a) There are 10 to 20 non-rainy days before the rainfall; b) The rainfall lasts more than one day with a total amount exceeding 50 mm; c) The surface runoff is almost zero during and after the rainfall process, which is estimated by the change of measured water level and the stage-storage curve.

The result rainfall events are shown in Table 7.3 with an average total amount of 90.4 mm, indicating that the conditions are optimal for the antecedent soil moisture without surface runoff following non-rainy days (W_m). It is also found that the water level rises at an average rate of 5.3 mm per day during six days after the rainfall events. Therefore, based on the speculated subsurface runoff volume from previous soil moisture, R_b is 5.3 mm, W_{Lm} is 31.8 mm (5.3×6), and W_{Um} is 58.6 mm (90.4-31.8). After choosing the closest meteorological stations in the China Daily Ground Climate Dataset, is regarded as 4.2 mm by averaging the evaporation consumption on non-rainy days from June to September between 2006 and 2012. Thus, regardless of how much it rains, the remaining soil moisture in the upper layer gradually evaporates over the next 14 days (58.6 / 4.2), and N is finally obtained. Combined with the basic information of the Hengshan Reservoir, the values of the model parameters are shown in the last column of Table 2.

Table 7.3 Typical rainfall events for model parameter acquisition

Date of rainfall process	Total rainfall amount (mm)	Estimated surface runoff volume (mm)	Number of non-rainy days before	Number of non-rainy days after
2006.08.08-2006.08.11	105.2	1.3	11	10
2006.09.09-2006.09.15	102.1	2.1	12	27
2007.07.07-2007.07.16	93.2	3.3	17	9
2009.07.05-2009.07.09	91.2	0.0	13	7
2011.07.10	81.1	1.0	18	15
2011.09.05-2011.09.09	101.3	1.6	14	10
2012.03.18-2012.03.22	81.0	0.2	13	26
2013.07.04	74.4	1.4	9	11
2014.08.26-2014.08.28	83.7	1.7	15	13

Aiming at intensive rainfall in the flood season, the forecasting model is to extend information for reservoir flood discharge and ensure the dam safety. Meanwhile, according to the precipitation magnitude provided by the State Flood Control and Drought Relief Headquarters, the 24-hour total precipitation is classified as heavy rain (25 to 50 mm), rainstorm (50 to 100 mm), heavy rainstorm (100 to 200 mm), etc. (Zhang and Li, 1992). Therefore, based on the actual flood discharging logs from 2006 to 2014, 106 days whose total rainfall amount exceeds 25 mm are chosen to verify the model's accuracy. There includes 75 days of heavy rain, 27 days of rainstorm, and 4 days of heavy rainstorm. Since the sample days are discrete, the verification is performed by comparing the measured water levels on the next day with forecasted water levels. The latter are based on actual changes of storage capacity and stage-storage curve of the Hengshan Reservoir. The analysis contains the following three folds:

(1) Based on the scatter diagram of forecasted and measured water levels on the above samples, a linear regression model is established (Fig. 7.8). According to the result values of slope and R^2, the two water levels generally coincide with each other, which indicates that forecasted values are close to the measurements.

Fig. 7.8 Comparison between forecasted and measured water levels

(2) The forecast errors are analysed by dividing the samples into three classes based on the precipitation magnitude (Fig. 7.9). As the average errors and error distributions shown in the Fig., the forecasted values are more likely to be higher than measurements in the heavy rain class, while in the rainstorm and heavy rainstorm class, the results are the opposite.

Fig. 7.9 Analysis of the forecast error according to the precipitation magnitude

(3) Due to the slight water level variations in the Hengshan Reservoir, the absolute errors are used and analysed by logarithmic, exponential and power regression with the total rainfall amount of the above samples (Fig. 7.10A; 7.10B; 7.10C). The result R^2 values are approximately 0.4, which implies that the amount can explain the forecast errors to some degree but not completely. The regression coefficients are significantly negative and pass the t-test at 99% confidence level in all the three models; thus, the forecast errors gradually decrease as the total rainfall amount increases, as shown by the curves. Meanwhile, their F-values are significantly positive at 99% confidence level, indicating that the three models have also passed the F-test and have equal variances. So, the three regressions are similar in nature, and the above conclusion is robust and reliable regardless of the model type.

According to the national standard of hydrological forecasting (MWR, 2008), the accepted range of water level forecast errors is within 20% of measured

Fig. 7.10 Analysis of the water level absolute error and daily rainfall amount. Three types of regression models are created in (A), (B) and (C), while the verification result of the forecasting model is shown in (D).

variations, and takes 0.1 m as a minimum value. Thus, 0.1 m is regarded as the maximum absolute error in this study. Based on the above results, although there is some error for the heavy rain class, the forecast is acceptable for rainfall events, whose daily amount is above or close to the heavy rainstorm level (Fig. 7.10D).

7.3.2 Application to flood control management

Following the previously described methodology, the WebGIS-based system is implemented and applied in the Hengshan Reservoir. The system functions are presented from the perspective of a reservoir manager, and a real example when Typhoon Chan-hom hit southeastern China is used to verify the decision-support for flood discharge during intensive rainfall in the flood season.

The manager first retrieves the basic information about the target reservoir, including the current water level and the storage capacity, the highest and lowest water level in history, and recent discharging logs (Fig. 7.11A). In the reservoir map controls, a geographic base map, a 2.5D map and panoramic views are provided to help explore the entire reservoir area. The manager could retrieve the daily and hourly data in statistical charts, and compare the water levels in a sketch map. During the continuous rainfall in the flood season, the manager switches the system into the flood discharging mode (Fig. 7.11B). The forecasting model is executed by integrating the 24-hour total rainfall amount, the current water level, and a particular discharging scheme. When the measured water level is acquired on the next day, the manager could analyse the accuracy and optimize the model parameters online.

When Typhoon Chan-hom arrived on July 11th, 2015, the water level was 33.86 m. Based on the 24-hour total rainfall amount of 117.2 mm and the

Fig. 7.11 Web interfaces of the flood control management system, which include reservoir routine management in (A), and decision-support for flood discharge in (B).

antecedent soil moisture estimation, the forecasted inflow flood volume would be 126.5 million m^3, and the water level would rise to 35.42 m until the next day (Fig. 7.12). Since it was around the middle of the flood season, the water level could not exceed the FLWL, so flood discharging decisions must be made immediately. Based on the concise and deterministic data, the manager loaded and compared three discharging schemes, which would respectively reduce the water level to 33.84 m, 33.90 m, and 33.94 m in the next 24, 12 and 20 hours. Due to the minimal effect on the downstream channels and the reserved time for post-discharging modifications, the last scheme with a total outflow discharge of 150 m^3/s was selected. The measured water level on the next day was 33.98 m with an absolute error of 0.04 m, which was within the above-mentioned error threshold; thus, the system effectively helps the reservoir manager to formulate appropriate discharging schemes, so that the water level and capacity is maintained in a safe range during intensive rainfall in the flood season.

Fig. 7.12 A real example of flood level forecasting and discharging schemes comparison during heavy rainstorm

7.3.3 Discussion

The numerous and widely distributed small reservoirs are featured by the uneven levels of management and limited hydrological measurements. Their storage capacity is normally maximized for multiple purposes, but discharged for flood control and conservation during the flood season (Rodrigues et al., 2012;

Song et al., 2015). In such circumstances, this study presents the WebGIS-based system as an effective tool to guide the flood control management, particularly for reservoir managers. It is different from the existing flood early warning systems for large river basins and reservoirs (Roo et al., 2011; Wu et al., 2014; Thiemig et al., 2015), which focus on flood routing simulation for different lead times, nor the flood assessment systems for local areas (Shivakoti et al., 2011; Demir and Krajewski, 2013; Horita et al., 2015), which currently adopt web-based technologies to involve different communities. Since several institutional and social conditions have to be fulfilled when applying sophisticated flood control systems (Cools et al., 2016), a targeted and adaptable tool could play a more useful role in practical flood discharge of small reservoirs.

In order to make the result system concise and practical, this study introduces the agile development method and tools during the entire system development process (Fig. 7.2). They help to embody multidisciplinary knowledge from different fields, such as reservoir management, flood discharge, hydrological modelling, map development and software integration. Since different experts usually prefer popular methods within their respective domains, a loosely coupled structure of different tools and languages is used (Fig. 7.3). Besides, the reservoir map browse is composed of three independent widgets, considering the uncertainties of spatial materials possessed by different reservoirs.

As the core of the system, the forecasting model is mainly based on rainfall in the catchment and the water level in front of the dam. Aiming at concise information for flood discharge, it is a simplified transformation of lumped conceptual hydrological models, and adopts a horizontal water level to generalize the fluctuating water surface caused by flood routing (Zhao, 1992; Shi et al., 2009; Lü et al., 2013). Meanwhile, the model currently uses daily lead time, considering the actual formulation mode of flood discharging scheme in the study area, and the applicability to other similar reservoirs that possess only daily hydrological measurements. A shorter forecast range, e.g., hourly, is also

Hydro-Environmental Assessment of Small Water Bodies:
From Local to Global Scales

available using this method, but the corresponding flood discharge may have to include several results to make an overall decision.

According to the verification based on historical rainfall events, the forecast errors gradually decrease as the total rainfall amount increases (Fig. 7.10D). In addition, the forecasted water levels are more likely to be greater than measurements in lighter rainfall, but smaller during heavier rainfall (Fig. 7.9). The explanation and analysis are as follows:

(1) When acquiring the parameter values of the study area, typical rainfall events are chosen based on the empirical principles that surface runoff doesn't form during and after the rainfall process, but the water level gradually rises in the following days. Then, the total rainfall amount of each typical event is averaged to estimate the value of W_m; however, the evaporation consumption is ignored during the calculation process. Therefore, the result soil moisture value is slightly over-estimated, which is presumably the reason why the forecasted inflow flood volume tends to be smaller during heavier rainfall.

(2) Runoff usually forms quickly after heavy rainfall, owing to the saturated soil moisture in the flood season. However, during the non-flood season, the rainfall infiltration from the upper to the lower soil layer requires further refinement, and the actual evaporation consumption should be greater than the flood season (Vieux, 2001; Lü et al., 2013). So the forecasted runoff volume tends to be greater than the actual value. This explains why the forecasted water level is more likely to be greater after lighter rainfall.

Therefore, detailed evaporation analysis is recommended for further improvement of the model. When estimating the antecedent soil moisture, the evaporation consumption in the upper layer should be classified by months and weather conditions, and deducted from the value of W_m. In order to acquire more accurate forecasts during the non-flood season and lighter rainfall, a bit more complex hydrological models are worth trying when distributed characteristics with relatively high-resolution of the catchment are available. For example, the applicable CN values of the SCS-CN model could be calibrated combined with water level variations, so that the result runoff volumes are comparable (Mishra and Singh, 2003; Shi et al., 2009). Besides, the usage of the WFlow model with rainfall interception and kinematic wave modelling could give the runoff process for each stream channel.

Due to the relatively sufficient flood discharging capacity and water storage requirements of the study area, the system currently relies on occurred rainfall, rather than numerical weather predictions, such as the Weather Research and the Forecasting model (Skamarock et al., 2005; Bartholmes and Todini, 2005), and satellite- and radar-based imagery data (Park and Hur, 2012; Wu et al., 2014). But meteorological forecasts are important to provide additional lead time to flood control preparedness, especially for the small reservoirs affected by flash floods. In addition, user interfaces of the water level comparison are used to help formulate discharging schemes. This method is concise, but still requires empirical judgement by reservoir managers. Based on the above runoff process simulation, stage-wise flood control operation rules and optimization algorithms currently used in cascade reservoirs and river-reservoir systems (Chou et al., 2015; Che et al., 2015), have good reference value to balance flood prevention and water storage, as well as to flood discharge of small reservoirs.

7.4 Conclusions

Small reservoirs and their current status of flood management are garnering increased attention from both researchers and decision-makers. In this study, a WebGIS-based flood control management system with a single targeted forecasting model is outlined to provide decision support for flood discharge during intensive rainfall in the flood season. The features and contributions are summarized as follows:

(1) In order to make the result concise and effective for reservoir managers, agile development methods are adopted during the entire implementation process, which helps to combine multidisciplinary knowledge from different fields. A loosely coupled structure of different tools and languages is used to

integrate reservoir map browse, flood level forecasts, and discharging schemes comparison, which enables the experts to use their popular methods in different domains.

(2) Based on the hydrological measurements of rainfall in the catchment and the water level in front of the dam, a flood level forecasting model with daily lead time is established by estimating the antecedent soil moisture and the accumulated inflow flood volume. The forecast result is acceptable for the rainfall amount above or close to the heavy rainstorm level, according to the national standard of hydrological forecasting of China.

Currently applied in the Hengshan Reservoir in the lower reaches of the Yangtze River, the system is validated by historical rainfall events and a trial operation during the typhoon season. It is featured by the usage of limited hydrological data that a small reservoir possesses, and the user-friendly interfaces for routine management and flood discharge for reservoir managers, which make it adaptable to other small reservoirs in humid regions. Since the optimization for flood control management is a complex and multifaceted issue, further improvements to the flood level forecasting model and study on the stage-wise flood control operation are recommended to better support the flood discharge of small reservoirs.

References

Alfieri, L., Burek, P., Dutra, E., et al., 2013. "GloFAS-global ensemble streamflow forecasting and flood early warning." *Hydrol. Earth Syst. Sci.* 17 (3): 1161-1175.

Bartholmes, J. and Todini, E., 2005. "Coupling meteorological and hydrological models for flood forecasting." *Hydrol. Earth Syst. Sci. Discuss.* 9 (4): 333-346.

Brhel, M., Meth, H., Maedche, A., et al., 2015. "Exploring principles of user-centered agile software development: A literature review." *Inform. Software Tech.* 61: 163-181.

Chang, F., and Chang, Y., 2006. "Adaptive neuro-fuzzy inference system for prediction of water level in reservoir." *Adv. Water Resour.* 29 (1): 1-10.

Che, D., and Mays, L. W., 2015. "Development of an optimization/simulation model for real-time flood-control operation of river-reservoirs systems." *Water Resour. Manag.* 29 (11): 3987-4005.

Cheng, C., and Chau, K., 2004. "Flood control management system for reservoirs. Environ." *Model. Softw.* 19(12): 1141-1150.

Chou, F. N. F., and Wu, C. W., 2015. "Stage-wise optimizing operating rules for flood control in a multi-purpose reservoir." *J. Hydrol.* 521: 245-260.

Ciarapica, L., and Todini, E., 2002. "Topkapi: A model for the representation of the rainfall-runoff process at different scales." *Hydrol. Process.* 16 (2): 207-229.

Cools, J., Innocenti, D., O'Brien, S., 2016. "Lessons from flood early warning systems. Environ." *Sci. Policy* 58: 117-122.

Demir, I., and Krajewski, W. F., 2013. "Towards an integrated flood information system: centralized data access, analysis, and visualization. Environ." *Model. Softw.* 50: 77-84.

Horita, F. E. A., de Albuquerque, J. P. D., Degrossi, L. C., et al., 2015. "Development of a spatial decision support system for flood risk management in Brazil that combines volunteered geographic information with wireless sensor networks." *Comp. Geosci.* 80: 84-94.

Lü, H., Hou, T., Horton, R., et al., 2013. "The streamflow estimation using the Xinanjiang rainfall runoff model and dual state-parameter estimation method." *J. Hydrol.* 480: 102-114.

Mishra, S. K., and Singh, V. P., 2003. "SCS-CN method." In *Soil Conservation Service Curve Number (SCS-CN) Methodology.* Dordrecht: Springer Verlag.

MWR 2008. Standard for Hydrological Information and Hydrological Forecasting (GB/T22482-2008). The Ministry of Water Resources, MWR of the People's Republic of China.

Park, J. H., and Hur, Y. T., 2012. "Development and application of GIS based K-DRUM for flood runoff simulation using radar rainfall." *J. Hydro-*

environ. Res. 6 (3): 209-219.

Rodrigues, L. N., Sano, E. E., Steenhuis, T. S., et al., 2012. "Estimation of small reservoir storage capacities with remote sensing in the Brazilian Savannah region." *Water Resour. Manag.* 26 (4): 873-882.

Roo, A. D., Thielen, J., Salamon, P., et al., 2011. "Quality control, validation and user feedback of the European Flood Alert System (EFAS)." *Int. J. Digit. Earth.* 4: 77-90.

Shi, Z., Chen, L., Fang, N., et al., 2009. "Research on the SCS-CN initial abstraction ratio using rainfall-runoff event analysis in the three Gorges area, China." *Catena* 77 (1): 1-7.

Shivakoti, B. R., Fujii, S., Tanaka, S., et al., 2011. "Remote sensing and GIS application for river runoff and water quality modeling in a hilly forested watershed of Japan." *J. Hydroinform.* 13 (2): 198-216.

Skamarock, W. C., Klemp, J. B., Dudhia, J., et al., 2005. "A description of the advanced research WRF version 2." DOI:10.5065/D68S4MVH.

Song, W., Jiang, Y., Lei, X., et al., 2015. "Annual runoff and flood regime trend analysis and the relation with reservoirs in the Sanchahe River Basin, China." *Quatern. Int.* 380-381: 197-206.

Thiemig, V., Bisselink, B., Pappenberger, F., et al., 2015. "A pan-African medium-range ensemble flood forecast system." *Hydrol. Earth Syst. Sci.* 19 (8): 3365-3385.

Van Der Knijff, J. M., Younis, J., De Roo, A. P. J., 2010. "LISFLOOD: A GIS-based distributed model for river basin scale water balance and flood simulation." *Int. J. Geogr. Inf. Sci.* 24 (2): 189-212.

Vieux, B. E., 2001. *Distributed Hydrologic Modeling Using GIS.* Dordrecht: Springer Verlag.

Werner, M., Cranston, M., Harrison, T., et al., 2009. "Recent developments in operational flood forecasting in England, Wales and Scotland." *Meteorol. Appl.* 16 (1): 13-22.

Werner, M., Schellekens, J., Gijsbers, P., et al., 2013. "The Delft-FEWS flow forecasting system." *Environ. Model. Softw.* 40: 65-77.

Wu, H., Adler, R. F., Tian, Y., et al., 2014. "Real-time global flood estimation using satellite-based precipitation and a coupled land surface and routing model." *Water Resour. Res.* 50(3): 2693-2717.

Zhang, Y., and Li, X. W., 1992. Flood control manual. The State Flood Control and Drought Relief Headquarters, Beijing, China.

Zhao, R. J., 1992. "The Xinanjiang model applied in China." *J. Hydrol.* 135 (1-4): 371-381.

Zhao, X. Y., Wang, Z. S., Sheng, J. B., et al., 2014. "Statistics of small dam failure and result classified analysis." *China Water Resources* 10: 33-35.

Chapter

⑧

Developing an Integrated 2D & 3D WebGIS Platform for Scattered Points of Interest

8.1 Introduction

Small, scattered points of interest, such as various natural resources including small water bodies, are ubiquitous worldwide. Landslide hazards of various types and scales is a type of typical small, scattered points of interest from large scale and database management perspectives. Increasing local rainfall extremes, as well as human activities are leading to more frequent landslide occurrence, resulting in losses of both property and human life (Petley, 2012). On the basis of field surveys, regional databases have been developed to improve our understanding of landslide hazards and opportunities for mitigation (Damm and Klose, 2015; Guzzetti et al., 2012). Meanwhile, the precondition, preparatory and triggering factors have been studied to support landslide prevention, early warning and emergency response (Arnone et al., 2011; Kapucu, 2008; Baum and Godt, 2010). In this process, software-based management platforms coupled with modern GIS techniques, and enhanced by interactive graphical user interfaces, are widely adopted and proved to be a synthetic method to combine and utilize the various types of data and information, so that local governments, social organizations and the general public can be better involved in the risk reduction efforts (Mansourian et al., 2006; Mantovani et al., 2010; Zhang et al., 2011; Assilzadeh et al., 2010; Salvati et al., 2009; Wu et al., 2014).

Landslide database is the basic but most important asset in such an applicable platform, as it not only gives insight into the distribution and characteristics of past hazards, but also provides materials for the susceptibility and risk assessment (Foster et al., 2012; Guzzetti et al., 2012; Galli et al., 2008). Over the past two decades, there has been considerable progress in the development of landslide databases across the world. Some official products are managed by national or state geological surveys, government departments, and their equivalents (Colombo et al., 2005; Damm and Klose, 2015; Pennington et al., 2015; Van Den et al., 2012). Other research catalogs, such as CRED and ICL databases, are compiled based on media reports and online data sources by non-government organizations (CRED, 2011; Kirschbaum et al., 2015; Zhang et al., 2011). Although most of the databases are updated at least once a year or after major events, active and regular coordination and synchronization is seldom carried out among their hosts and offices, owing to the different reasons for creation and maintenance (Mansourian et al., 2006; Trigila et al., 2010; Galli et al., 2008; Andersson-Sköld et al., 2013). The independent management procedure brings inconsistency and redundancy to local regions when analyzing hazard situations based upon different datasets. On the other hand, it is difficult to share and reuse the latest information that is essential to the mitigation policies and measures.

WebGIS, centered on web services and presented in web pages, is the mainstream of GIS development. It focuses on real-time sharing of geo-spatial data and models on the internet, as well as reuse and extension of functional services (Kawasaki et al., 2012; Fu and Sun, 2010). WebGIS is commonly used as an integrative method for landslide skim, query and analysis based on compiled databases. For example, Mantovani et al. (2010) provided an open-source solution to publish the landslide geomorphological maps, so that road maintenance was improved in Olvera, Spain. Thiebes et al. (2013) integrated a limit-equilibrium model into a web-based GIS to facilitate the slope stability analysis which requires varieties of input data. Huang et al. (2015) proposed a web-based platform combined with a wireless sensor network for automatic debris-flow monitoring in the area hardest-hit by the Wenchuan earthquake. On the basis of detailed survey and monitoring data, these platforms provide a feasible evaluation and prediction environment for local landslide-prone areas. However, it is an elaborate and time-consuming task to collect detailed information for large hilly and mountainous regions, especially when relying on different government departments (Scolobig and Pelling, 2016; Pennington et al., 2015; Dai et al., 2002). WebGIS-based techniques are supposed to continuously improve the compiled database by integrating multiple data services, and to provide latest references to both routine management and emergency response following sudden landslide incidents.

Landslide database and related geological and climatic conditions are widely assembled into 2D base maps (Trigila et al., 2010; Salvati et al., 2009). In order to provide users with an immersive visualization and analysis environment, 3D technologies are gradually taken into consideration. Dai et al. (2014) presented a parallel 3D model based on the smoothed particle hydrodynamics to simulate landslide motion across complex 3D terrain. Huang et al. (2015) used the virtual environment of Microsoft Bing Maps to visualize the monitoring devices and their real-time signals for debris-flow early warning. Travelletti and Malet (2012) integrated multi-source and multi-resolution data into a 3D geometrical structure for mudslide stability analysis and hydro-mechanical modeling. The use of 3D technologies makes the professional knowledge in landslide database, such as magnitude, failure mechanisms and affected areas, more comprehensible to decision-makers and the common people. However, due to inherent issues in the user experience, such as coordinate uncertainty of the clicked position, and line-of-sight blocked by 3D terrain (Del et al., 2007), 3D cannot completely replace 2D during the analysis of a hazard situation (Liu et al., 2014; Thum and De, 2015). A current topic for landslide platform designers and users is therefore how can 2D maps and 3D scenes be used together to generate a superior display and analytic environment in WebGIS.

The goal of this paper is to present a novel 2D and 3D WebGIS-based platform for landslide management and emergency response. It tries to solve two main research questions: (1) What kind of platform can integrate different data services, including detailed geological surveys and other useful materials, so that landslide database can be easily kept up-to-date by the joint effort of different government departments? (2) What methodology can couple 2D WebGIS, which plays a role in hazard query and statistics in routine management, with 3D WebGIS, which provides a useful tool for terrain analysis and sketch map plotting during emergencies? The main objective of this platform is to effectively support local authorities to manage and respond to landslide hazards during the long-term risk reduction process, to provide an easy-access environment to help avoid the risks for the general public, and finally to make contributions to geo-hazard prevention and mitigation. The paper first introduces Zhejiang Province as the study area, where landslide incidents are frequent and widespread. Next, based on the analysis of multi-level management and emergency response in large hilly and mountainous regions, the scalable architecture, pre-defined modules, and three key algorithms are illustrated. The paper then presents several application results to verify the functions and capabilities of the platform, and finally concludes by discussing the technical characteristics and further recommendations when performing the web-based landslide information management.

8.2 Study area

Located in the southeast coastal region of China, Zhejiang Province has a continental area of 101,800 km^2, and a population of 70 million. Zhejiang's step-like terrain is mostly comprised of fault-cutting hills and mountains, inclined from the southwest towards the northeast. Landscapes are diverse within this area, including hills, river valleys, and coastal plains. Zhejiang belongs to the subtropical monsoon climate zone with annual precipitation of 1,200~2,200 mm. There are two rainy seasons each year. The first appears from early May to late June, and is characterized by the low intensity and long duration rainfall events. The latter is in the typhoon season from late July to September, which brings heavy rainstorms along the coast (Gang and Gang, 2014). As one of the most economically developed provinces in China, human activities like deforestation, construction and slope cultivation are quite active, which make the surface unstable and vulnerable to landslide occurrence. So, the area is ideal for studying landslides in China, owing to the specific geological environment, geomorphic characteristics, climatic settings, and human activities.

From 2000 to 2013, landslide field surveys have covered all the 62 counties in the hilly and mountainous regions of Zhejiang Province (Department of Land and Resources of Zhejiang Province, 2008-2013). The resulting database currently contains 5,329 hazards and potential sites, of which about 68% are shallow soil slides, 19% rock falls, and 13% debris-flows. Statistics reveal that the three types of hazards exist in every city (Fig. 8.1A), and cover 66,700 km^2 with more than 700 villages affected. Among these hazards, about 88% occur between May and September, especially in June, July and August, which is consistent with the trend of mean monthly rainfall (Fig. 8.1B). From 2011 to 2013, a total of 2,022 landslides occurred and killed 11 people, causing a direct economic loss of 187 million yuan (Department of Land and Resources of Zhejiang Province, 2008-2013). Landslide hazards in the study area are recognized to be widespread and mostly of small size, but may result in serious consequences during heavy rainfall.

Fig. 8.1 The landslide statistics by city (A) and month (B). The three types of landslides in (A) contain both occurred hazards and potential sites. The four scales of occurred hazards in (B) are compared with mean monthly rainfall in the past 20 years from (Gang and Gang, 2014).

8.3 Methodology

8.3.1 Requirement analysis

Since the International Decade for Natural Disaster Reduction (IDNDR, 1990~2000) launched by the United Nations, many structural and nonstructural measures are gradually carried out in landslide-prone areas worldwide. Taking Zhejiang Province as an example, there contains the susceptibility mapping and emergency planning based on field surveys, the rainfall monitoring for entire regions, the development of rainfall thresholds for early warning, and the displacement monitoring for potential sites (Li et al., 2011; Zhang et al., 2011; Ma et al., 2015; Wu et al., 2014). These measures achieved remarkable results, but it is still difficult to avoid landslide incidents. In order to collect and manage the increasing data and information, an integrated software platform is proposed by the local authorities. It is not the replacement of the above measures, but a supplementary method to ensure the latest and effective references to landslide routine management and emergency response. It is also expected to publish the prevention information for the awareness and education of the common people.

Landslide multi-level management is a typical mode for large hilly and mountainous regions. Since the hazards are geographically scattered and difficult to predict, their management is mostly comprised of several areas of responsibility, which are usually consistent with the administrative levels and regions (Scolobig and Pelling, 2016; Dai et al., 2002; Sharma et al., 2012). The multi-level specifically contains provincial, municipal and county levels in Zhejiang Province, and in each level, there are one or several administrative regions, which are parallel to each other. After a careful investigation, the main responsibilities of the three levels are shown in Fig. 8.2. Since landslide field surveys and risk reduction projects are carried out yearly, only the cooperation and simultaneous improvements from these levels and regions can keep the compiled database consistent and up-to-date. Besides, results from other prevention measures, such as the susceptibility maps, have to be integrated and presented online. Therefore, flexible and scalable architecture is essential for the platform during the long-term and iterative process of implementing the above responsibilities for different administrative levels.

Fig. 8.2 The framework of landslide multi-level management. The green and red dots respectively mark the responsibilities during routines and emergencies.

Emergency response is the procedure to minimize casualties and property losses during and after a sudden landslide incident. Although the rescue team and auxiliary equipment may differ from place to place, it claims immediate decisions and actions after receiving the occurrence report (Assilzadeh et al., 2010; Bhatia, 2017; Kaku and Held, 2013). The overall procedure of landslide emergency response in Zhejiang Province is shown in Fig. 8.3, which involves all the administrative regions in the three levels where a hazard occurs. In order to support consultation and command, as well as on-the-spot rescue and evacuation, all useful materials related to the hazard site have to be gathered as soon as possible. Therefore, the platform needs landslide fast location identification in both the 2D map and the 3D scene, and quick access to the background information and rainfall situations nearby. Useful tools like map measurement and sketch map plotting are also indispensable to help improve the pre-set emergency plans in the light of the actual conditions nearby.

Fig. 8.3 The sketch map of landslide emergency response

In a previous management platform, designers make efforts to cope with several landslide-related government departments when acquiring supplementary data, such as geographic base maps and precipitation (Colombo et al., 2005; Baum and Godt, 2010). The local and independent storage pattern always results in data inconsistency and information out-of-date. Fortunately, the development of web technologies has enabled the implementation and maintenance of various data services from these departments (Kawasaki et al., 2012; Fu and Sun, 2010; Yue et al., 2007), such as Bureau of Surveying and Mapping, and Bureau of Meteorology in China. The platform is supposed to obtain and integrate these online services, so that the designers and users can be easily provided with the latest supplementary data.

8.3.2 Architecture design

Architecture design is the core of the software platform and plays a decisive role in accomplishing the requirements and targets (Hofmeister et al., 2000). Network architecture and software architecture designed for this platform embody the consideration for the network level of platform users in the future in addition to suitable development components and their relationship to balance functions, user-experience and extendibility.

The platform is hierarchically published to intranets (private networks of local authorities) in order to support multi-level management and emergency response in large hilly and mountainous regions. The logic view of the network architecture is in a tree-like structure (Fig. 8.4), and the leaves could expand according to the specific administrative regions in municipal and county levels. In each system, the functions are fully deployed, but the corresponding data, e.g., the landslide records, only contains information within the administrative region. Thus, the platform naturally possesses the authorization capabilities. In county level systems, hazard managers save the survey data in their local databases after field investigation, and then upload the changes to the municipal system. In municipal and provincial level systems, managers can perform advanced analyses based on the landslide databases they possess. Additionally, the platform is published to the Internet (public network) only in the provincial level via the Internet web server, so that the common people can easily acquire the latest prevention information.

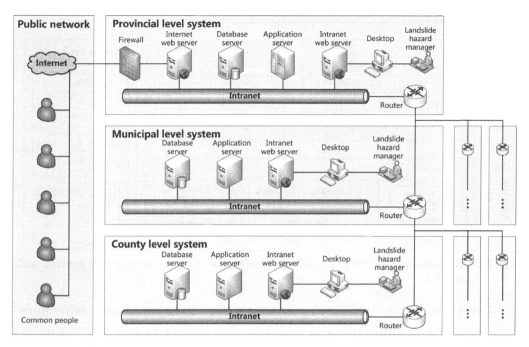

Fig. 8.4 The network architecture. The platform deploys hierarchically in intranets for provincial, municipal, and county levels, while publishes to the Internet only in the provincial level.

The platform adopts a three-tier software architecture (data tier, service tier and application tier) enhanced by remote services from other government departments (Fig. 8.5). In every system of the three levels, the three tiers are fully deployed. The data tier is in the database server. It stores landslide survey data and susceptibility maps in Microsoft SQL Server 2008, and saves prevention documents and multimedia data in a Windows file system. The service tier is in the application server. It contains spatial and non-spatial data services in Microsoft IIS 8. ArcGIS Server 10 is used as a publishing tool for susceptibility maps and an online geoprocessing environment for rainfall data. The model-view-controller (MVC) web application built on NHibernate, Spring. Net, and Asp.Net takes charge of non-spatial data services from the tier below. The application tier is loaded in the intranet web server for hazard managers in

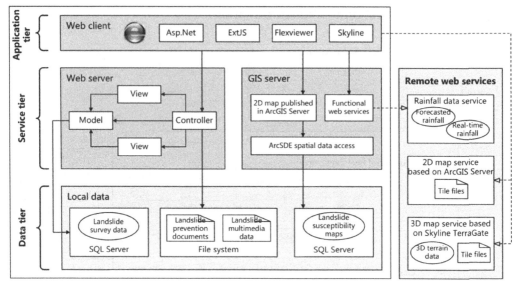

Fig. 8.5 The software architecture, which contains three tiers within and remote web services.

the three levels. It adopts Ext JS and Apache Flex to enhance user experience. Geographic and thematic map display is in ArcGIS Flexviewer, while 3D terrain and rainfall analysis are in Skyline ActiveX component. In this software architecture, several local and remote web services are integrated in the application tier, which helps to solve the issues of multi-source data acquisition.

8.3.3 Modules design

Based on the above analysis and design, the functions of the platform are organized into three main modules, which are landslide data management, emergency response and information publishing (Fig. 8.6). Since most of the functions cover the three tiers in the software architecture, they are relevant to the database server, application server and web server in the network architecture. Their details are as follows:

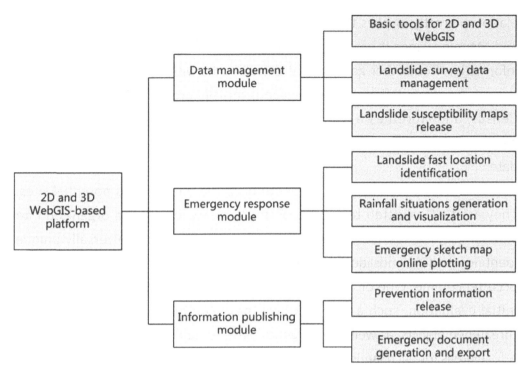

Fig. 8.6 The modules and main functions

8.3.3.1 Data management module

As a basis of the platform, this module stores, maintains and displays all the landslide-related spatial and non-spatial data. It extends the compile database by integrating several web services and a 2D and 3D combined WebGIS, so that the survey data improvement and decision-making of different administrative levels have a consistent and effective environment online.

In order to reveal the location and surrounding conditions of a landslide hazard from multiple viewpoints, three map display modes are adopted. They are 2D map only, 3D scene only, and 2D and 3D associated display. In the last mode, the 2D map widget and the 3D scene widget are automatically linked to display the same area. Besides, basic tools such as map pan, measurement, feature identification, and layer control are also provided. The geographic base map contains several kinds of background information, such as geomorphic features, road conditions, village distribution, resident population, etc. The platform directly invokes the data services established by Bureau of Surveying and Mapping, since it has the thematic data on good authority by aggregating information from other government departments with expertise. For example, the resident population of each village is initially produced by Bureau of the Census, and stored as a thematic layer on the base map. In this way, the background information in the module is credible and automatically kept up-to-date.

Landslide survey data contains both occurred hazards and potential sites. They are first collected by field experts in county levels, and organized into standardized survey tables, including core attributes, like universally unique identifier (UUID), landslide type, geographical coordinates, location description, occurrence time, type of the sliding body, state of activity, etc. In addition, the initial evaluation, such as the risk level based on potential magnitudes and threatened objects, as well as emergency plans containing evacuation routes and refuge areas for each landslide record are produced as illustrations, and stored in the module. Based on the above data records, shallow soil slides, rock falls, and debris-flows are dynamically rendered in 2D map and 3D scene by the specific symbols (Fig. 8.7) to reveal their location and distribution. Since survey data is stored in different systems of the three levels, managers in each system are only authorized to process the data within their administrative region, and obligated to upload the changes to the direct superordinate and subordinate systems, so that landslide information is consistent and effective for the entire regions.

Shallow soil slide   Rock fall   Debris-flow

Potential   Potential   Potential
shallow soil slide   rock fall   debris-flow

Fig. 8.7
Landslide hazard symbols, which have a clockwise direction and all point to the north here.

Since the susceptibility maps are typical results from existing prevention measures, the platform doesn't contain the assessment process, but provides landslide records as input data, and an online release environment via web services for the assessment experts in provincial and municipal levels. Based on the above architecture, decision-makers could easily retrieve the susceptibility level of the places in their administrative region, and check the feasibility of planned activities, such as road construction, routine patrol and preventive resettlement by map overlaying analysis.

8.3.3.2 Emergency response module

Since sudden landslide incidents are difficult to completely avoid, this module helps to minimize the casualties and property losses after receiving an occurrence report. It provides fast access to the background information and rainfall situations nearby to facilitate the decision-making and on-the-spot rescue actions. Landslide location identification is the first step in the emergency procedure. During an urgent situation, local authorities may receive one of the four kinds of occurrence reports. Specially, they are:

(1) Coordinates of the hazard site. In this case, the occurrence is first found and reported by field experts on patrol during the rainy season. It can be directly located by the reported longitude, latitude, landslide type and direction;

(2) Description of the affected area. The hazard is reported by local residents before an expert arrives, and only descriptive information is available owing to their limited professional knowledge. The hazard can be roughly located by the place name and corresponding coordinates stored in the platform, and further identified according to their additional description and geomorphic features presented in 3D scene;

(3) Description of the landmarks nearby. The situation and location is similar to (2);

(4) Structured landslide data record. It happens only after the subordinate system handling the report and submitting upwards. Since the record contains

basic information of the hazard site, the location is similar to (1).

After location identification, real-time rainfall and forecasted rainfall are obtained and visualized via the data services established by Bureau of Meteorology. It is not the early warning, but the latest rainfall information to help grasp the on-the-spot situations during emergencies. Real-time rainfall is calculated and displayed by clicking on a particular rainfall station near the landslide site (Fig. 8.12B). The point feature layer of the stations is generated by the metadata of web service, including station ID, name and coordinates, and published in 3D scene beforehand. Forecasted rainfall is produced and organized as hypsometric contours by automatically downloading the data packets and choosing a specific future period (Fig. 8.12C).

Besides, sketch map plotting tools are provided to help improve and visualize the emergency plans in this module. For example, if the reported hazard is not covered by the field survey and landslide database, decision-makers can mark evacuation routes and temporary refuge areas online based on the surrounding geomorphic features and construction distribution presented by 2D and 3D WebGIS. These materials are important references to on-the-spot rescue actions.

8.3.3.3 Information publishing module

This module is designed to utilize landslide-related information outside the intranets of local authorities. It contains prevention information for the common people, and emergency report for on-the-spot rescue actions. The former is based on a news website deployed in the provincial level system and published to the Internet (Fig. 8.5). The latter quickly generates a document containing location, surrounding conditions, and emergency sketch map of the sudden landslide incident. It is implemented by inserting the descriptions and screenshots during the emergency response procedure into a template file of Microsoft.

8.3.4 Algorithms for 2D and 3D dynamic display

In a WebGIS-based platform, Flexviewer is widely adopted as a 2D map widget, while Skyline is popular as a 3D scene widget (Tate et al., 2011; Wang et al., 2014). Although the two components provide plenty of basic functions, they are actually separated from each other when simultaneously used. In order to generate a combined environment for landslide management, and to use external data services to support emergency response, the enhancements for 2D and 3D dynamic display are carried out in the following three aspects.

8.3.4.1 Landslide hazard symbols rendering

As a basic language element of GIS, map symbols are the main visualization tool to represent geo-spatial information. The point feature in 2D map is commonly used to describe landslide distribution (Colombo et al., 2005; Guzzetti et al., 2012; Salvati et al., 2009), and several line types are used to indicate the profile of a landslide area (Trigila et al., 2010; Pennington et al., 2015). In order to vividly represent shallow soil slides, rock falls and debris-flows in large prone areas, the above-mentioned hazard symbols (Fig. 8.7) are dynamically rendered in a point feature layer in both 2D and 3D WebGIS.

Based on the landslide type, coordinates and direction from the compiled database, the image symbol is placed in the 2D map widget with a specific turning-angle (Fig. 8.8A), while its size is decided by the current map scale. If the plane image is directly moved upwards in 3D scene, it may be partially obscured by 3D terrain meshes. Therefore, a method for dip angle calculation is used to make the image symbol parallel with the terrain above (Fig. 8.8). First, as in 2D map, determine the image position and turning-angle based on the data record, and decide the size by the current viewpoint height. Second, project the four corner points of the image vertically upwards to the DEM terrain surface; obtain the X, Y and Z coordinates of each projected point by inverse distance weighted (IDW) interpolating (Allen et al., 2011) on its four surrounding DEM points; and then calculate the dip angle of the projected rectangle. Finally, move the image symbol to a specific distance above the terrain surface (e.g., 100m in this platform), and render it by the result dip angle in 3D scene.

Fig. 8.8 Landslide hazard symbols rendering in the 2D map and the 3D scene. The image symbol with a turning-angle at the bottom of (A) shows the result in 2D map; after the dip angle calculation and image movement, it is parallel with the 3D terrain in the top of (B); a dynamic rendering example based on landslide records is presented in (C).

8.3.4.2 2D and 3D associated display

Most traditional GIS applications are based on 2D maps, which are essentially an abstract symbol system. Technical advances allow for the integration of elevation information into the basic data model in 3D GIS to generate more sophisticated virtual exploration, terrain analysis, 3D modeling, etc. (Liu et al., 2014; Travelletti and Malet, 2012; Thum and De, 2015; Dai et al., 2014). In this platform, 2D and 3D WebGIS are associated, which means the 2D map widget and 3D scene widget are linked and automatically display the same area to help fast concentrate on landslide incidents in an urgent situation (Fig. 8.12A). The key method is to establish an event trigger mechanism to match the 2D display scale and the 3D viewpoint height, as well as their display extent. Specifically, to control the 3D scene from the 2D map widget, the event handler on display extent change is presented as follows:

(1) Obtain the coordinates of the center point and map scale after display extent changes in the 2D map.

(2) Acquire the 3D viewpoint height by the 2D map scale based on scale-height value pairs prepared in the platform.

(3) Invoke the flight method of the 3D scene, and pass three input parameters, which are the coordinates of center point, the result viewpoint height, and a vertical downward view angle.

The method for a 2D map responding a 3D scene is similar, but the event handler should be disabled before display extent changes and enabled afterwards to avoid an endless loop. Due to the geodetic coordinate system that 2D WebGIS usually adopts, and the geographic coordinate system, such as WGS84 that 3D WebGIS mostly uses, there are slight deviations during the association, although it doesn't affect the hazard location identification and situation analysis.

8.3.4.3 Rainfall contours online plotting

Among the triggering factors, rainfall is an important indicator to predict landslide occurrence and its development trend. So based on the forecasted data from Bureau of Meteorology, hypsometric rainfall contours are generated and plotted online in 3D WebGIS, so that decision-makers could grasp the rainfall situations in a timely manner to guide on-the-spot rescue and evacuation. The technical roadmap includes three main steps: offline preprocessing, online geoprocessing, and online plotting (Fig. 8.9).

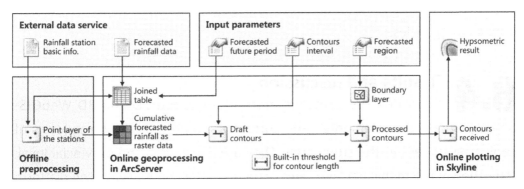

Fig. 8.9 Technical roadmap of forecasted rainfall contours online generation and plotting

In the offline preprocessing stage, the point feature layer of rainfall stations is generated and published in 3D WebGIS. The online geoprocessing remotely obtains the forecasted rainfall data, and receives three input parameters, which are the forecasted future period, the contours interval, and the forecasted regions. The details are as follows:

(1) The forecasted data records according to the input future period are extracted from regular downloaded data packets, and associated with the point layer of rainfall stations by a multi-table join operation. Based on the IDW interpolation tool in ArcToolbox, cumulative forecasted rainfall is generated in a raster file, and converted into contours.

(2) To make the draft contours simple and intuitive, some post-processing is necessary. The clip tool in ArcToolbox is used to remove the external contours by overlaying the forecasted region. Besides, the length of each contour is calculated in order to remove the ones that are shorter than the built-in threshold (e.g., 10km in this platform).

(3) The result contours are smoothed and exported to the application tier.

The above procedures are implemented by ArcGIS ModelBuilder (Allen, 2011) and published as an online geoprocessing service in ArcServer in the service tier. Invoked by Skyline ActiveX component in the application tier, it plots the result contours in 3D WebGIS (Fig. 8.12C). This method gives an example of integrating external data services with internal geoprocessing services, so that the latest supplementary data can be dynamically interpreted and visualized during emergencies.

8.4 Results and discussion

Following the previous methodology, the 2D and 3D WebGIS-based platform is implemented and currently accepted by the three levels of geological surveys in the study area. They are Zhejiang in the provincial level, Hangzhou in the municipal level, and Chun'an in the county level. The platform provides decision-makers with a more effective tool to perform landslide multi-level management and emergency response in large prone areas. It also helps to acquire the prevention information for the common people.

8.4.1 Landslide multi-level management

Landslide multi-level management contains routine work such as hazard query, statistical analysis, survey data update and improvement. In the platform, hazard query can be performed by determined attribute values, or using point, line, or area spatial selection tool in 2D WebGIS. Hazard statistics help to plot pie charts and histograms for local administrative regions. A hypothetical hazard manager in Hangzhou City would apply the platform in his routine work as follows:

To grasp the rock fall situation in Chun'an County of Hangzhou City, he inputs the query extent and landslide type, and then acquires the rock fall distribution in 2D WebGIS (Fig. 8.10A). Brief information of a certain rock fall is instantly presented in a pop-up panel when the image symbol in the 2D map or the data record in the result table list is clicked. In addition, the detailed information, such as the standardized survey table, emergency and prevention plans, as well as multimedia data about this hazard can be easily retrieved. The initial evaluation, including the threatened objects nearby, is stored as illustrations in the multimedia data, rather than presented in 2D map, since the map is a direct reference to the data service maintained by other government departments with expertise. With the help of spatial statistics tools, the pie chart for three counties of Hangzhou City is plotted (Fig. 8.10B), and the proportions of different landslide types are acquired. After overlaying the susceptibility map on a 2D geographic base map (Fig. 8.10C), the susceptibility level of each village within the administrative region is checked, which helps to formulate patrol plans during the rainy season.

Fig. 8.10 Application results of landslide multi-level management, which contains hazard query (A), spatial statistics (B), susceptibility map display (C), and standardized survey data submission and verification (D).

After annual field surveys, he receives new data records of potential hazards from the Chun'an County system, and merges them into the local database after verification. The standardized landslide survey table, taking shallow soil slide as an example (Fig. 8.10D), contains detailed information such as the occurrence time, location descriptions, geological and hydrological environments, historical deformations, inducing factors, the state of activity, etc.

It is an effective method to try a system in limited areas, and then to carry out optimizations and promotions step-by-step in other places when performing landslide information management (Mansourian et al., 2006;

Assilzadeh et al., 2010). According to the proposed architecture and modules, other cities and counties in Zhejiang Province could rapidly implement their systems and join into the whole platform. In addition, based on the joint effort from three levels of geological surveys, as well as the expertise of other government departments, landslide-related information is kept consistent and up-to-date in the platform.

8.4.2 Landslide emergency response

In spite of annual field surveys and real-time monitoring measures, sudden landslide incidents are inevitable. In order to minimize the losses, emergency response collects the nearby conditions as soon as possible, so that field investigation and rescue actions could be well-founded. Based on the consistent and latest landslide-related information, this work is enhanced by the 2D and 3D WebGIS-based platform. Below, this paper presents a real debris-flow scenario in Chun'an County in Aug. 2011 to verify the decision support capabilities during emergencies.

(1) Occurrence report reception. During the rainy season, decision-makers in the provincial level receive a phone call from the patrolling expert that a large debris-flow is in progress on the north of L.F. Temple in Chun'an County, and it threatens 200 households nearby. The emergency response procedures are launched immediately.

(2) Hazard location identification. Since the debris-flow site is covered by previous field surveys, it is quickly located in both 2D map and 3D scene based on the data record (Fig. 8.11A). So, the nearby road conditions and village distribution are acquired. Decision-makers believe that the urgent situation requires the field working team to help evacuate the villagers immediately, and

Fig. 8.11 Application results of landslide emergency response. There is hazard fast location identification (A), real-time rainfall analysis (B), forecasted rainfall contours online generation (C), and sketch map plotting (D).

the information collecting to keep on.

(3) Rainfall analysis. In order to acquire the weather conditions near the debris-flow site, decision-makers load the rainfall stations in 3D scene, and click the nearest L.K.K. station to obtain the real-time rainfall information (Fig. 8.11B). The cumulative rainfall from nearly 1 hour to 24 hours indicates that it is a typical rainfall induced hazard. To estimate its further development trend, forecasted rainfall contours till the next 24 hours are generated and plotted online (Fig. 8.11C). Unfortunately, it is 150 mm nearby, which shows that the local area would be suffering from rainstorm, and the rescue work is in urgent.

(4) Rescue and evacuation sketch map plotting. Since the survey data doesn't contain emergency plans for the debris-flow site, decision-makers have to plot the sketch map just based on the information presented by 2D and 3D WebGIS, as well as their expertise and experience. With the help of online plotting tools, the key points of their ideas and plans are easily expressed in 3D scene, which contain the debris-flow catchment, danger zone, evacuation routes and temporary refuge areas (Fig. 8.11D). The sketch map is the basis of further decisions and actions.

(5) Information export. The above materials, including background information, real-time and forecasted rainfall situations, as well as the sketch map, are collected and saved into a document. It is then sent to the field working team to provide integrative spatial information support for their rescue and evacuation actions.

8.4.3 Prevention information for public access

Landslide prevention information is published to the Internet based on a standalone news website deployed in the provincial level system. It contains four columns, which are prevention plans, landslide cases, hazards nearby, and prevention knowledge (Fig. 8.12A). Particularly, the last two columns are closely related to the common people who live in large hilly and mountainous regions. If occurred hazards or potential sites in the landslide database are close to a village, they are first marked in a static distribution map based on offline preparations, and then released with a brief warning message and the photos taken nearby. The task is continuously performed after annual field surveys and regional risk assessments by hazard managers in the provincial level, so that all the landslide records will be checked. Besides, special topics and training materials are regularly produced as videos and pictures to help develop the awareness of landslide risks, as well as the prevention and first-aid skills for the villagers (Fig. 8.12B). As a separate extension of the 2D and 3D WebGIS-based platform, the website makes the latest prevention information accessible and comprehensible to the common people with basic Internet conditions and limited professional knowledge.

Fig. 8.12 The prevention information for public access. The hazard distribution near J.L. village is showed in (A), and an instance of prevention knowledge is presented in (B).

8.4.4 Discussion

Landslide databases of various spatial scales and levels of detail are the basis of risk reduction measures. However, active and regular coordination and synchronization is seldom carried out among their hosts and offices, which brings inconsistency and redundancy to local regions when analyzing hazard situations based upon different datasets (Mansourian et al., 2006; Trigila et al., 2010; Galli et al., 2008; Andersson-Sköld et al., 2013). In such circumstances, this paper proposes the integrated WebGIS-based platform as an effective tool for local authorities to better perform landslide routine management and emergency response. It is different from existing platforms that provide a feasible evaluation and prediction environment for local areas where detailed survey and monitoring data is available (Thiebes et al., 2013; Mantovani et al., 2010; Huang et al., 2015), but presents as a collaborative environment for geological surveys and other government departments with expertise to keep the compiled database consistent and up-to-date during the long-term risk reduction efforts.

Since landslide multi-level management is a typical mode for large hilly and mountainous regions, the platform adopts a scalable network architecture (Fig. 8.4) and pre-defined modules (Fig. 8.6) for government departments in different administrative levels and regions, so that branch systems could be rapidly implemented. Instead of elaborating the customized settings, such as data partitioning, user privileges, and detailed functions in each system, this paper focuses on the two main characteristics of the platform. One is the integration of web services to maintain the latest landslide-related information; another is the combined utilization of 2D and 3D WebGIS during routines and emergencies.

As one of the major issues impacting current WebGIS development, web service integration is vital to the interoperation of multi-source and heterogeneous spatial data (Kawasaki et al., 2012; Fu and Sun, 2010). In this platform, the technology is widely applied in the application tier (Fig. 8.6). For example, the acquisition of the geographic base map and the association between 2D map and 3D scene directly invoke the services from Bureau of Surveying and Mapping, while the interpretation and visualization of rainfall situations rely on the internal geoprocessing services and external meteorological services from Bureau of Meteorology. These methods not only free the designers from basic data storage and access, but also help the users to acquire the latest information. However, it is worth mentioning that related functions are sometimes unstable, owing to unexpected changes in the application program interfaces (APIs) of external web services. So, the balance between efficiency and reliability needs to be reconsidered when using external services in the short term, while in the long term, it is worthwhile as they are fully prepared.

The platform combines 3D WebGIS with 2D WebGIS, and takes advantages of their respective superiorities in dynamic display and situation analysis. The 2D map widget and 3D scene widget are associated during map extent change and hazard location identification, so that both distribution and concentration on particular landslide hazards can be fast achieved. The proposed landslide symbol rendering method vividly represents their type, location and direction, while the images in 3D scene are parallel and close to the terrain surface below (Fig. 9). However, it is difficult to reveal the profiles of unstable slopes and the catchments of debris-flows using the point-based method (Colombo et al., 2005; Guzzetti et al., 2012; Salvati et al., 2009). On the basis of detailed field survey data, spatial characteristics of landslide hazards are supposed to be extracted via data templates or semantic models (Liu et al., 2014; Dong et al., 2013), so that they can be marked more naturally and accurately by several line types in both 2D and 3D WebGIS.

The result platform is currently accepted by three geological surveys, which are Zhejiang in the provincial level, Hangzhou in the municipal level, and Chun'an in the county level. In order to take effect in all the cities and counties in the province, and provide references to other landslide-prone areas, both technical improvement and training programs have to be taken into account. When the platform is applied in more administrative regions, design for load and concurrency is required (Hofmeister et al., 2000). For example, the network and database have to handle simultaneous requests from several subordinate systems. On the other hand, the user expertise is one of the decisive factors for the application and effectiveness of a software-based management method (Scolobig and Pelling, 2016; Pennington et al., 2015). So, the training programs are indispensable for decision-makers to widely accept and understand the web-based cooperative procedures and spatial information support capabilities of the platform.

In the current version, a standalone website of landslide prevention information is designed for public access. Although hazard managers in the provincial level work meticulously to keep villagers informed of the nearby hazards, this method mostly relies on their active access to the web pages. Since a timely warning is usually more important than the expressions of landslide risks, especially in emergencies (Bhatia, 2013; Kapucu, 2008), another important enhancement in the future versions is to build mobile clients as a portable reminder (Bhattacharya et al., 2012). Based on real-time positions provided by the mobile phones, the enhancement can not only change the information pull mode into the push mode, but also offer customized announcements to different stakeholders.

8.5 Conclusion

As typical small, scattered points of interest, landslide hazards of various types and scales are a ubiquitous problem, especially in large hilly and mountainous regions. In this study, an integrated 2D and 3D WebGIS-based platform is outlined for landslide routine management performed by several government departments, and emergency response following sudden hazard occurrence. The scalable network architecture and three-tier software architecture are designed based on requirement analysis of multi-level management, fast spatial decision support for emergencies, and web service integration. Subsequently, three main modules are proposed, and key algorithms for 2D and 3D dynamic display are elaborated. There include hazard symbols rendering, the association between 2D map and 3D scene, as well as online interpretation and visualization of forecasted rainfall data.

The user-friendly platform is currently accepted by the three levels of geological surveys in Zhejiang Province, China, and plays an important role in landslide database improvement during the long-term risk reduction process. The web service integration has proved effective for keeping landslide-related information consistent and up-to-date, since it changes the independent management procedure and the isolated storage pattern into collaborative effort from different government departments with expertise. The combined utilization of 2D and 3D WebGIS takes advantage of their respective superiorities in analysis and display. The 2D map is mainly for hazard query, spatial statistics and map overlaying analysis during routines, while the 3D scene is useful in terrain exploration, rainfall situation analysis and sketch map plotting in emergencies.

Both technical improvement for the platform and training programs for decision-makers are highlighted when performing the software-based management method. The former contains enhancement in architecture for concurrent access and optimization for hazard symbols rendering, while the latter aims at a wide accept and understand of the web-based functions and capabilities. Another important recommendation is to build mobile clients for timely warnings of landslide risks to the common people. In spite of these further improvements, the platform and technologies have provided good references to landslide prevention and mitigation in large landslide-prone areas.

References

Allen, D.W., 2011. *Getting to know ArcGIS ModelBuilder.* Redlands: ESRI Press.

Andersson-Sköld, Y., Bergman, R., Johansson, M., et al., 2013. "Landslide risk management: A brief overview and example from Sweden of current situation and climate change." *Int. J. Disast. Risk Re.* 3: 44-61.

Arnone, E., Noto, L.V., Lepore, C., et al., 2011. "Physically-based and distributed approach to analyze rainfall-triggered landslides at watershed scale." *Geomorphology* 133(3): 121-131.

Assilzadeh, H., Levy, J.K., Wang, X., 2010. "Landslide catastrophes and disaster risk reduction: A GIS framework for landslide prevention and management." *Remote Sens.* 2(9): 2259-2273.

Baum, R.L., and Godt, J.W., 2010. "Early warning of rainfall-induced shallow landslides and debris flows in the USA." *Landslides* 7(3): 259-272.

Bhattacharya, D.,Ghosh, J.K., Samadhiya, N.K., 2012. *Development of a geo-hazard warning communication system: Geo-referenced dissemination of real time warnings through the mobile telecom network.* Koln: LAP Lambert Academic Publishing.

Colombo, A., Lanteri L., Ramasco, M., et al., 2005. "Systematic GIS-based landslide inventory as the first step for effective landslide-hazard management." *Landslides* 2(4): 291-301.

CRED, EM-DAT, 2011. *The OFDA/CRED International Disaster Database, Centre for Research on Epidemiology of Disasters-CRED.* Brussels: Université Catholique de Louvain.

Dai, F.C., Lee C.F., Ngai Y. Y., 2002. "Landslide risk assessment and management: An overview." *Eng. Geol.* 64(1): 65-87.

Dai, Z., Huang, Y., Cheng, H., et al., 2014. "3D numerical modeling using smoothed particle hydrodynamics of flow-like landslide propagation triggered by the 2008 Wenchuan earthquake." *Eng. Geol.* 180: 21-33.

Damm, B., and Klose, M., 2015. "The landslide database for Germany: Closing the gap at national level." *Geomorphology* 249: 82-93.

Del, F.V., Paolino, L., Pittarello, F., 2007. "A usability-driven approach to the development of a 3D web-GIS environment." *Journal of Visual Languages & Computing* 18(3): 280-314.

Department of Land and Resources of Zhejiang Province, Zhejiang Geological Environment Bulletins, Zhejiang, China, 2008-2013.

Dong, S., Wang, X., Yin, H., et al., 2013. "Semantic enhanced WebGIS approach to visualize Chinese historical natural hazards." *J. Cult. Herit.* 14(3): 181-189.

Foster, C.,Pennington, C.V.L., Culshaw, M.G., et al., 2012. "The national landslide database of Great Britain: development, evolution and applications." *Environ. Earth Sci.* 66(3): 941-953.

Fu, P., and Sun, J., 2010. *Web, GIS: principles and applications.* Redlands: ESRI Press.

Galli, M., Ardizzone, F., Cardinali, M., et al., 2008. "Comparing landslide inventory maps." *Geomorphology* 94(3): 268-289.

Gang, L., and Gang, L., 2014. "Analysis on the spatial-temporal characteristics of precipitation variation tendency in Zhejiang Province." *Subtropical Soil & Water Conservation* 26(3): 9-14.

Guzzetti, F., Mondini, A.C., Cardinali, M., et al., 2012. "Landslide inventory maps: New tools for an old problem." *Earth-Sci Rev.* 112(1): 42-66.

Hofmeister, C., Nord, R., Soni, D., 2000. *Applied software architecture*. New York: Addison-Wesley Professional.

Huang, J., Huang, R., Ju, N., et al., 2015. "3D WebGIS-based platform for debris flow early warning: A case study." *Eng. Geol.* 197: 57-66.

Kaku, K., and Held, A., 2011. "Sentinel Asia: A space-based disaster management support system in the Asia-Pacific region." *Int. J. Disast. Risk Re.* 6: 1-17.

Kapucu, N., 2008. "Collaborative emergency management: better community organising, better public preparedness and response." *Disasters* 32(2): 239-262.

Kawasaki, A., Berman, M.L., Guan, W., 2012. "The growing role of web-based geospatial technology in disaster response and support." *Disasters* 37(2): 201-221.

Kirschbaum, D., Stanley, T., Zhou, Y., 2015. "Spatial and temporal analysis of a global landslide catalog." *Geomorphology* 249(15): 4-15.

Li, C., Ma, T., Zhu, X., et al., 2011. "The power-law relationship between landslide occurrence and rainfall level." *Geomorphology* 130(3): 221-229.

Liu, J., Tang, H., Zhang, J., et al., 2014. "Glass landslide: the 3D visualization makes study of landslide transparent and virtualized." *Environ. Earth Sci.* 72(10): 3847-3856.

Ma, T., Li, C., Lu, Z., et al., 2015. "Rainfall intensity-duration thresholds for the initiation of landslides in Zhejiang Province, southeast China." *Geomorphology* 245(15): 193-206.

Mansourian A., Rajabifard A., Zoej M.V., et al., 2006. "Using SDI and web-based system to facilitate disaster management." *Comput. Geosci.* 32(3): 303-315.

Mantovani, F., Gracia, F.J., Cosmo, de, et al., 2010. "A new approach to landslide geomorphological mapping using the open source software in the Olvera area (Cadiz, Spain)." *Landslides* 7(1): 69-74.

Pennington, C., Freeborough, K., Dashwood, C., et al., 2015. "The national landslide database of Great Britain: Acquisition, communication and the role of social media." *Geomorphology* 249(15): 44-51.

Petley, D., 2015. "Global patterns of loss of life from landslides." *Geology* 40(10): 927-930.

Salvati, P., Balducci, V., Bianchi, C., et al., 2009. "A WebGIS for the dissemination of information on historical landslides and floods in Umbria, Italy." *Geoinformatica* 13(3): 305-322.

Scolobig, A., and Pelling, M., 2016. "The co-production of risk from a natural hazards perspective: Science and policy interaction for landslide risk management in Italy." *Nat. Hazards* 81(1): 7-25.

Sharma, U., Scolobig, A., Patt A., 2012. "The effects of decentralization on the production and use of risk assessment: Insights from landslide management in India and Italy." *Nat. Hazards* 64(2): 1357-1371.

Tate, E., Burton, C.G., M. Berry, et al., 2011. "Integrated hazards mapping tool." *Transactions in GIS* 15(5): 689-706.

Thiebes, B., Bell, R., Glade, T., et al., 2013. "A WebGIS decision-support system for slope stability based on limit-equilibrium modelling." *Eng. Geol.* 158: 109-118.

Thum, L., and De, P.R., 2015. "2D and 3D GIS-based geological and geomechanical survey during tunnel excavation." *Eng. Geol.* 192: 19-25.

Travelletti, J., and Malet, J.P., 2012. "Characterization of the 3D geometry of flow-like landslides: A methodology based on the integration of heterogeneous multi-source data." *Eng. Geol.* 128: 30-48.

Trigila, A., Iadanza, C., Spizzichino, D., 2010. "Quality assessment of the Italian Landslide Inventory using GIS processing." *Landslides* 7(4): 455-470.

Van, D. M., and Hervás, J., 2012. "State of the art of national landslide databases in Europe and their potential for assessing landslide susceptibility, hazard and risk." *Geomorphology* 139-140: 545-558.

Wang, T., Gao, Z., Ning, J., 2014. "Research on construction of web 3D-GIS based on Skyline." In *Remote Sensing and Modeling of Ecosystems for Sustainability XI.* DOI:92210I.1-92210I.5.

Wu, Y., Chen, L., Cheng, C., et al., 2014. "GIS-based landslide hazard predicting system and its real-time test during a typhoon, Zhejiang Province, Southeast China." *Eng. Geol.* 175: 9-21.

Yue, P., Di, L., Yang, W., et al., 2007. "Semantics-based automatic composition of geospatial web service chains." *Comput. Geosci.*33(5): 649-665.

Zhang, G., Chen, L., Dong, Z., 2011. "Real-time warning system of regional landslides supported by WEBGIS and its application in Zhejiang province, China." *Procedia Earth and Planetary Science* 2: 247-254.

Chapter
⑨

Exploring Global Collaboration Opportunities for Non-floodplain Wetland Conservation

9.1 Introduction

Wetlands provide multiple benefits and services that are essential in achieving the global Sustainable Development Goals (SDGs). Non-floodplain wetlands (NFWs) are a common type of wetlands, surrounded by uplands outside of floodplains and riparian areas (Mushet et al., 2015). These waters are numerous, widely distributed, and tightly linked to socio-economic development, but their typically small size (even down to 0.01 ha; Lane and D'Amico, 2016) and shallow nature (water depth < 6 m; Ramsar Convention, 1971) leave them frequently unmonitored and unmapped. Due to this, NFWs remain poorly protected and vulnerable to degradation and destruction caused by both hydro-climatic and anthropogenic drivers (e.g., agricultural intensification, urban expansion, eutrophication, salinization, and invasive species) (Cohen et al., 2016; Creed et al., 2017; Golden et al., 2019). Over the last decade, NFWs have however been increasingly recognized for their profound role in providing multiple ecosystem services, that were once primarily associated with larger, more widely studied systems such as coastal floodplain wetlands and lakes (Cohen et al., 2016; Rains et al., 2016; Golden et al., 2017; Lane et al., 2018). This has led to a range of recent initiatives highlighting the importance of NFWs and calling for collaborative efforts at national and transnational levels to achieve their sustainable use (Hill et al.,

2018; Sullivan et al., 2019; Cheng et al., 2020; Sayer et al., 2020; Swartz et al., 2021; Lane et al., 2022).

The Ramsar Convention is by far the most far-reaching international agreement on wetland conservation and sustainable use. A major objective of the Convention is to provide opportunities for the global community to learn and collaborate based on the geographical, functional, and biological representation of the Ramsar wetland sites (Bridgewater and Kim, 2021). This is yet far from being achieved for the world's NFWs, with lacking coordinated research and collaborative utilization at cross-regional to global scales. Starting with the terminology, "NFW" comes from the U.S. and has been derived from "isolated wetlands" and "geographically isolated wetlands (GIWs)" in recent years (Leibowitz and Nadeau, 2003; Mushet et al., 2015). Although the new term has gradually gained acceptance in academia for emphasizing the geographical location rather than falsely generalizing the hydrological isolation of these waters, it is used inconsistently and often accompanied by "GIWs", even in the recent research literature and government documents. While in other parts of the world, terms including (but not limited to) "small water bodies" (Biggs et al., 2017), "neglected freshwater habitats" (Hunter et al., 2017), "temporary wetlands" (Calhoun et al., 2017) and "wetlandscapes" (Thorslund et al., 2017; Ghajarnia et al., 2020), are used or partly used to study similar small vulnerable waters, with a different focus on wetland attributes (e.g., size, perimeter-area-volume relationship, and hydroperiod), functions (e.g., flood attenuation, nutrient retention, and biodiversity support), and study scales (e.g., individual wetlands, wetlands across landscapes, and wetlands at watershed and regional scales), respectively. Such marked heterogeneity has been cross-validated by several reviews and calls for improved research and collaborative utilization of these wetland systems (Hunter et al., 2017; Chen et al., 2019; Golden et al., 2019; Sayer et al., 2020), which indicate the prevailing perspectives at local and regional scales — NFWs are studied with different emphases under different motivations, depending upon where they are located and what we are interested in. The lack of further refinement and organization on the heterogeneity obscures the representativeness of these wetland systems and is an obstacle to collaborative research and management at larger scales.

Given these knowledge gaps, we raise three questions that are central for promoting the theory and practice of the sustainable NFW use to the global scale: (i) how extensively have NFWs been studied in different parts of the world? (ii) what are the patterns of their research focuses across different regions? and (iii) how can current research efforts aid in improving collaborative research and management of NFWs? We address these questions through the use of meta-analyses of recent scientific literature, an in-depth investigation of representative regions, and interdisciplinary research recommendations, with detailed methods and intermediate results in Appendix Section 8.1 to 8.4. Although any term has inherent limitations for phenomenon that actually spans a continuum (e.g., river vs. stream; Richardson et al., 2022), "NFW" is favored over the others here due to the essential characteristics that distinguish these waters from floodplain wetlands, and the willingness to remain neutral to those different research focuses and motivations. Our exploratory research can synthesize dispersed knowledge and galvanize a common conversation for NFWs in the global community. It can also aid management and protection for similar small vulnerable waters like ephemeral streams in the context of global freshwater challenges.

9.2 NFWs have been widely studied with strong North America biases

Using a dual-step search procedure, our meta-analysis on NFW research resulted in 2213 peer-reviewed articles, published over the last 20 years (2001-2020) and cataloged in the Web of Science™ Core Collection. This procedure contains a first step to retriece three synthesis types of studies (i.e., review, commentary, and perspective articles), and then a second step to examine all published articles on NFWs, based on keyword combinations as informed by results from the first step. Throughout the discovery process, particular attention was paid to the abstract, figures, and conclusions of each retrieved article to ensure that the research object belongs to, contains, is within or related to NFWs, rather than floodplain wetlands. Specifically, for the papers that did not specify "non-floodplain" in the abstract and conclusions, a visual inspection of the figures was conducted on a distance criterion of 10-meters away from rivers, coasts, and floodplains. This distance criterion was used to identify putative GIWs from the U.S. Fish and Wildlife Service National Wetlands Inventory, relative to the U.S. Geological Survey National Hydrography Dataset, according to Lane and D'Amico (2016). We used the above literature set to map the proportion of NFWs being studied across different continents, after identifying specific and common wetland types and their research frequency. Detailed paper search methods, measures to ensure NFWs, and intermediate results are in Appendix Section 9.1.

We first identified 36 specific NFW types, and organized them into 9 main types and 27 subtypes, according to the inland wetland classification scheme of the Ramsar Convention (Ramsar Convention Secretariat, 2013) and the U.S. Environmental Protection Agency (Figure 9.1a). Ponds, including its subtypes: farm ponds, temporary ponds, multi-pond systems, karst ponds, chain-of-ponds, and Delmarva ponds, are most studied. Ponds account for 34% of the 2,213 articles, despite a recent functional definition involving the surface area, depth, and coverage of emergent vegetation, that distinguishes them from

Fig. 9.1 Specific and common NFW types with the number and proportion of articles that each type occurs (a), and proportions of main wetland types being studied on each continent (b), according to the 2,213 NFW research articles published over 2001-2020. The main NFW types have the same color as their subtypes in (a), and the same color as themselves in the doughnuts in (b). See Appendix Section 8.1 for further interpretations and uncertainty considerations.

lakes and wetlands (Richardson et al., 2022). Other NFW types, in descending order of research frequency (ranging from 27 to 5%), include pools, constructed wetlands, marshes, potholes, swamps, fens, and bogs (including their subtypes), while other less-studied wetland types aggregately rank third at 16%.

Based on the specific NFW types, relevant studies have been found across all the seven continents, but is dominantly located in North America (between 39% and 96%, for the 9 main types; Fig. 9.1b). This is presumably due to the original search term "NFW" used, and development of legislative/regulatory and management policies in the U.S. (Sullivan et al., 2020; Wade et al., 2022). Apart from potholes, in which the dominant subtype, i.e., prairie potholes, are endemic to North America, Asia, and Europe are the following continents with a combined share of 33-53% in NFW research, with research focuses on ponds, marshes, and constructed wetlands, each having a combined share close to or higher than that of North America. Africa, South America, and Australia show an average share of 3.3, 3.4, and 4.6%, respectively, with the total share of swamps (17%) close to that of Asia, while no studies on fens or bogs being found in Africa. Moreover, Antarctica has a small but noteworthy share at 0.9-1.9% for the categories of ponds, pools, and other wetland types. (i.e., 4 out of the 11 polar wetlands in Fig. 9.1a; see Table A8.2 for further elaboration). The geographic distribution of widely studied NFWs was cross-validated by several global wetland database, including HydroLAKES and MERIT Hydro, these datasets also highlight the inherent lack of a generalized framework for NFW research as they are both missing geographic distribution data for Antarctica (Hu et al., 2017; Zhang et al., 2021).

9.3 Emerging research patterns across representative NFW regions

The above meta-analysis revealed strong North America biases in the NFW studies, which we hypothesize stem from terminology rather than real geographical bias around existence and functionality. To confirm these biases, we selected 30 representative regions for an in-depth investigation of research characteristics across NFWs around the world (Fig. 9.2). The selection criteria were based around trying to include as much identified NFW types as possible, to cover a wide geographical distribution (by continent) and hydro-climatic settings (by the updated Köppen-Geiger climate classification system; Peel et al., 2007), and to cover active NFW study areas, based on recent literature and relevant projects on the sustainable wetland use. Further narrations on the selection criteria, grouping experiments, and analytical procedures are in Appendix Section 9.2. In particular, we analyzed the land-cover characteristics for grouped representative regions, and then linked these regions with a subset (229 site-related articles) of the above full set of articles to assess 12 research focuses (including hydrogeomorphology, water resource, hydrological and biogeochemical processes, nutrient retention, greenhouse gas regulation, biodiversity, ecosystem services, sustainable agriculture, inventory mapping, and climate change) on NFWs. These research focuses were summarized from the Millennium Ecosystem Assessment (WHO, 2006) and retrieved literature, and quantified to converge scientific and management concerns across the representative regions (Appendix Section 9.3).

Fig. 9.2 Investigated NFW representative regions (a) and their average proportions of land-cover types and research focuses, grouped by the five climate zones used (b). See Appendix Section 8.2 for selection criteria, wetland types, satellite imagery, and location information of the representative regions, and further interpretations of the asterisks and dot lines in the statistical chart.

Using the updated Köppen-Geiger climate classification system as a grouping criterion, and based on the grid-formatted Esri 2020 Global Land Cover map, our results identified similarities in land-cover types around NFWs in different regions of the world. Specifically, tropical and cold zones have common dominant land-cover types around NFWs, including trees, flood vegetation, and scrub-shrub (17%-28%), followed by a smaller proportion of water, grass, crops and build area (2%-15%), despite their huge difference in climatic conditions. Similarly, both arid and polar zones have a higher proportion of scrub-shrub and bare ground (17%-57%), although the second largest land-cover type in arid zones (20%), i.e., crops, doesn't exist in polar zones, while snow-ice in polar zones (26%) is statistically higher (p-value < 0.1) than that of all other zones combined and almost none in arid zones. Moreover, around NFWs, the proportion of crops, trees, and build area in temperate zones (38, 24, and 21%, respectively) and waters in cold zones (6%) are statistically higher than that of the same land-cover type in all other zones combined.

In contrast to the biased NFW studies, commonalities and patterns were discovered with regard to the 12 NFW research focuses. Biodiversity stands out from the others, with high prevalence across all climate zones (44% on average here and throughout). Hydrological and biogeochemical processes are also common across all zones (both 40%), whereas wetland protection is popular in tropical, arid, and temperate zones (52%), but with a lower focus in cold and polar zones (21%). Inventory mapping, nutrient retention (although a subset of biogeochemical processes), and sustainable agriculture have been least studied (13%-14%), although the first one in tropical zones and the latter two in temperate zones are more prominent. Additionally, evaluated publications appear to focus to a larger extent on climate change (44%) and greenhouse gas regulation (40%) in tropical, polar, and cold zones, and ecosystem services in tropical zones (38%) with moderate ecosystem service concerns in arid, temperate and cold zones (25%). Hydrogeomorphology in arid, cold and polar zones (26%) and water resources in arid and temperate zones (24%) are moderately studied.

These emerging Patterns research focal areas, discovered here at global scales, can be attributed to cross-regional differences in hydro-climatic settings and NFW land-cover characteristics. Specifically, greater motivations in studying hydrological and biogeochemical processes are presumably due to relatively limited precipitation but equally necessary freshwater resources in arid and cold zones (Lawford et al., 2013; Pekel et al., 2016). Higher sensitivities to climate change in tropical, polar, and cold zones at extreme latitudes (IPCC, 2021) have led to more attention towards climate change related topics, including greenhouse gas regulation. Extensive cropland and human activities are associated with all land-cover types except snow-ice, confirming higher prevalence of nutrient retention in temperate zones and ecosystem services in all zones outside the polar zones, respectively.

9.4 New opportunities for global coordinated management efforts

We have demonstrated that NFWs have been widely studied across all continents (including the often-overlooked Antarctica), but with strong North America biases. Moving beyond terminology to wetland functionality, we have further shown commonalities and patterns of research focuses in representative regions of different NFW types and climate zones, and highlighted their underappreciated value in the world's major economies. This scientific evidence implies that efforts to develop convergent knowledge and obviate the North

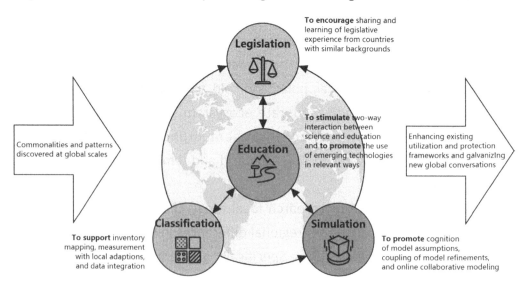

Fig. 9.3 Schematic diagram of four focus areas for promoting sustainable NFW use at global scales

America biases are feasible. We here present a vision for global collaboration, and outline four pathways that will enhance our understanding and sustainable use of these wetland systems across regions —— from classification and simulation as theoretical and technical bases, to improved legislative/regulative support via mutual learning of governments, and bidirectionally reinforced education and science (Fig. 9.3). These recommendations, particularly for countries with similar geo-climatic settings and socioeconomic development levels (due to better opportunities to learn from each other), can complement existing frameworks of NFW utilization and protection for governments and agencies (Calhoun et al., 2017; Golden et al., 2017; Hunter et al., 2017; Chen et al., 2019; Swartz and Miller, 2021), and have the potential to galvanize broader international collaborations on sustainable wetland use.

Classification. NFWs have heterogeneous naming criteria in different geographical settings, e.g., relating to hydroperiod for temporary ponds (Calhoun et al., 2017), landforms for karst ponds (Hill et al., 2018), structures for chain-of-ponds (Williams et al., 2020), and purpose for farm ponds (Takeuchi et al., 2016) (Fig. 9.1a). The semantic mediation among terms is a fundamental issue that urgently needs to be recognized for better understanding and inventorying across regions, as well as reducing North America biases. A targeted categorization can then be built upon the main wetland types and subtypes identified, with particular focus on the ecosystem services that characterize NFWs (Fig. 9.2b) and hydrogeomorphic settings that shape NFWs (e.g., classification systems of Tiner, 2003 and Dvorett et al., 2012), plus additional reference to historical classics that reveal how NFWs evolve (e.g., historical records in Gao et al., 2015 and Poschlod and Braun-Reichert, 2017). On this basis, mutual aid with local adaption considerations (i.e., taking local conditions into considerations), especially on the dominant hydro-climatic characteristics and landscape processes, e.g., the freeze-thaw processes for polar wetlands (Rains, 2011) and irrigation practices for multi-pond systems (Chen et al., 2020b), is preferred, when developing novel technologies, such as hyperspectral remote sensing (Wu et al., 2019), in detecting NFWs' presence and physical characteristics. Together, these could be more reliable than the distance criterion currently in use, whether it is 10 meters or larger. At continental and global scales, reassembling exemplary and sophisticated datasets like the U.S. National Wetlands Inventory and Pan-European High-Resolution Layers, while integrating observations of various research focuses (e.g., the six data categories of the integrated monitoring framework proposed by Chen et al., 2019), with regular automatic or semi-automatic updates (e.g., via Google Earth Engine Data Catalog), is recommended to reflect the impact of human activities in disparate regions and boost watershed and climate change sciences (Berrang et al., 2015; Golden et al., 2021).

Simulation. Recent advances have demonstrated that incorporating NFWs into watershed modeling (from watershed areas ranging from ~10 km^2 like Chen et al., 2020b to ~1 thousand km^2 like Evenson et al., 2016, Yeo et al., 2019 and Zeng and Chu, 2021, and to ~0.5 million km^2 like Rajib et al., 2020) can improve the accuracy of runoff and nutrient yield simulations. However, challenges remain in understanding the cumulative and comprehensive effects of these landscape mosaics (i.e., small but nonnegligible aquatic patches that embedded in various land-cover types, according to Mushet et al., 2019 and validated by Fig. 9.2b) at continental or global scales, given our general lack of understanding of their underlying hydrological processes and sensitivity to climate alteration. In terms of land-cover characteristics and research focuses that characterize NFWs (Fig. 9.2), the emerging, cross-regional differences we demonstrated confirm the diversity of current model refinements, e.g., hydrological connectivity for prairie potholes (Lane et al., 2018), phosphorus retention for farm ponds (Chen et al., 2019), and biodiversity support for vernal pools (Sullivan et al., 2019) in the cold, temperate and arid climate zones, respectively; while the commonalities we presented highlight the need and potential of mutual learning of model assumptions and coupling of model refinements, especially when assessing the coevolution of natural and human systems around NFWs at larger scales. Additionally, separate modeling and parameter estimation for typical study sites, followed by integration and comparison of simulation results, can help eliminate the issues of inconsistent model objectives and structures due to the lack of unified knowledge (i.e., epistemic uncertainty), but requires further calibration and fidelity assessment to reflect cross-regional discrepancies. Open web-distributed collaborative modeling frameworks, from implementation standards like OpenMI to service-oriented platforms like CSDMS and OpenGMS (Salas et al., 2020; Chen et al., 2020a), are a viable solution, especially in the trend of sharing and reuse for monitoring data.

Legislation/Regulation. In contrast to the universal research focuses across the globe (Fig. 9.2), environmental legislation and regulation for NFWs is uneven, like the biased proportions of these systems being studied across different continents (Fig. 9.1b). A few countries, such as England (Biggs et al., 2005), have more than three decades of direct experience in pond monitoring and restoration. Others countries, such as U.S., have different policies in the federal and state governments — The former have once enforced rules that put NFWs at risk but now imminently promote understanding and amendment, while the latter have taken responsibilities and enacted additional, protective regulations (Florida in particular), although some states have not done so (Sullivan et al., 2019; Sullivan et al., 2020; Wade et al., 2022). Additionally, laws and regulations that implement indirect protection have been emerging in a few countries over the past two decades. Examples include Japan's Satoyama Initiative that promotes the sustainable use of rural natural resources (Takeuchi et al., 2016), China's Lake/River Chief Mechanism that ensures the water governance and ecological integrity throughout river basins (Wang et al., 2019), and Kenya and Tanzania's Wildlife Act that improves the biodiversity of national nature reserves with relevance to the safety of large mammals (Cockerill and Hagerman, 2020). These cases confirm the value of NFWs in larger ecosystems (i.e., rural, water and wildlife ecosystems) and the effectiveness of indirect protection by enhancing existing legal frameworks relating to greater social concerns, although the reality on the ground may not be as effective as the legal framework (Xu et al., 2019). Structured analysis and extension of all these ongoing stories in conjunction with comprehensive valuation and compensation of ecosystem services is recommended to promote win-win management investment for NFW protection and restoration. This recommendation is particularly useful for countries with similar geo-climatic settings around NFWs (Fig. 9.2) and socio-economic development levels, since similar monitoring, evaluating and management strategies can be employed, according to Flörke et al., 2013 and Aguilar, 2020.

Education. The sustainable use of NFWs can be an excellent testbed for two-way interactions between science and education. For popular education and science popularization, cross-regional comparisons based on either formal lectures or field trips can enhance the understanding of the ecosystem services provided by NFWs, and facilitate the incorporation of their valuation into macroeconomic indicators. For research and professional education, establishing connections between historical legacies of NFW terminology (e.g., chain-of-ponds and multi-pond systems in Fig. 9.1), commonalities and patterns in the characterizing land-cover types and research focuses (Fig. 9.2), and geographical, historical, political, and cultural driving factors behind, can help understand our diverse world, including the major current issues such as climate change and resource scarcity. Meanwhile, technological advances are constantly making interdisciplinary knowledge comprehensible and participatory. Virtual reality and gamification, as a basis of the metaverse (Lee et al., 2021), for example, can be used to develop immersive, targeted exhibitions on NFWs in conjunction with iconic wildlife species (e.g., salamanders for vernal pools in the northeastern U.S.; Brooks, 2005) and surrounding land-cover types (e.g., woodlands, plantations, grasslands, farmlands, irrigated ponds, canals, etc. for the Satoyama landscape in Japan; Takeuchi et al., 2016), as have been applied in the online version of some prominent museums (Lee et al., 2020).

The popularization of high-definition cellphone cameras, fifth generation mobile network, and edge machine learning capabilities (i.e., techniques that achieve real-time processing on resource-constrained terminal devices of the Internet of Things, reducing reliance on the cloud network) can enhance crowdsourced data collection and real-time sharing and analysis on the ecological conditions of these small, vulnerable waters, if properly used to keep the applications attractive to the public participants (including local planners, citizens, farmers, travelers, etc. as summarized by Chen et al., 2019).

9.5 Connecting the disconnected

Better use and protection of wetlands across the world is an important challenge without a simple solution. The knowledge of NFWs, as a valuable socio-economic and environmental resource, has facilitated local and regional efforts on their sustainable use, but cross-regional to global collaborations are lacking, due to the lack of consistent terminology, differentiated research focuses, and heterogeneity at different scales. We have, however, shown here, by our meta-analysis and in-depth investigation, that there are multiple commonalities that emerge for NFW systems, with wide studies on each continent, despite strong North America biases in the literature. To confirm these biases, commonalities and patterns in NFW research focuses were explored across global climate zones, including biodiversity, hydrological and biogeochemical processes, and wetland protection as top research hotspots, a larger extent on climate change and greenhouse gas regulation in tropical, polar, and cold zones, and inventory mapping, nutrient retention, and sustainable agriculture being least studied. These findings suggest that there is more that unites disparate wetland research and management efforts than we might otherwise appreciate. They can also help us move forward with practical work, connecting pathways for new opportunities for the sustainable NFW use by the global community — From (i) classification to support inventory mapping, measurement with local adaptions, and data integration, to (ii) simulation to promote the cognition of different model assumptions, coupling of model refinements, and online collaborative modeling, (iii) sharing and learning of legislation/regulation experience from countries with similar backgrounds, and (iv) stimulated two-way interactions between science and education with the use of emerging technologies. These recommendations should be further explored to improve collaboration and global realization of sustainable wetland systems.

References

Aguilar, N., 2021. Green gross domestic product (Green GDP) and sustainable development. In *Reduced Inequalities. Encyclopedia of the UN Sustainable Development Goals*, edited by Leal Filho, W., Azul, A.M., Brandli, L., Lange Salvia, A., Özuyar, P.G., Wall, T., Berlin: Springer Cham.

Berrang, F. L., Pearce, T., and Ford, J. D., 2015. "Systematic review approaches for climate change adaptation research." *Reg. Environ. Change* (15): 755-769.

Biggs, J, Williams P, Whitfield, M., et al. 2005. "15 years of pond assessment in Britain: Results and lessons learned from the work of pond conservation." *Aquat. Conserv.* 15: 693-714.

Biggs, J, Von, F. S., Kelly, Q. M., et al. 2017. "The importance of small waterbodies for biodiversity and ecosystem services: Implications for policy makers." *Hydrobiologia* 793: 3-39.

Brooks, R. T., 2005. "A review of basin morphology and pool hydrology of isolated ponded wetlands: Implications for seasonal forest pools of the northeastern United States." *Wetl. Ecol. Manag.* 13(3): 335-348.

Calhoun, A. J., Mushet, D. M., Bell, K. P., et al., 2017. "Temporary wetlands: Challenges and solutions to conserving a 'disappearing' ecosystem." *Biol. Conserv.* 211: 3-11.

Chen, M., Voinov, A., Ames, D. P., et al. 2020a. "Position paper: Open web-distributed integrated geographic modelling and simulation to enable broader participation and applications." *Earth-Sci. Rev.* 207: 103223.

Chen, W., Nover, D., Yen, H., et al. 2020b. "Exploring the multi-scale hydrologic regulation of multi-pond systems in a humid agricultural catchment." *Water Res.*184: 115987.

Chen, W., He, B., Nover, D., et al. 2019. "Farm ponds in southern China: Challenges and solutions for conserving a neglected wetland ecosystem." *Sci. Total. Environ.* 659: 1322-1334.

Cheng, F. Y., Van Meter K. J., Byrnes, D. K., et al., 2020. "Maximizing US nitrate removal through wetland protection and restoration." *Nature* 588: 625-630.

Cockerill, K., and Hagerman, S., 2020. "Historical insights for understanding the emergence of community-based conservation in Kenya: International agendas, colonial legacies, and contested worldviews." *Ecol. Soc.* 25(2).

Cohen, M. J., Creed, I. F., Alexander, L., et al., 2016. "Do geographically isolated wetlands influence landscape functions." *P. Natl. Acad. Sci. USA.* 113(8): 1978-1986.

Creed, I. F., Lane, C. R., Serran, J. N., et al., 2017. "Enhancing protection for vulnerable waters." *Nat. Geosci.*10(11): 809-815.

Downing, J. A., 2010. "Emerging global role of small lakes and ponds: Little things mean a lot." *Limnetica* 29(29): 9-24.

Downing, J. A., Prairie, Y. T., Cole, J. J., et al., 2006. "The global abundance and size distribution of lakes, ponds, and impoundments." *Limnol. Oceanogr.* 51(5): 2388-2397.

Dvorett, D., Bidwell, J., Davis, C., et al., 2012. "Developing a hydrogeomorphic wetland inventory: Reclassifying national wetlands inventory polygons in geographic information systems." *Wetlands.* 32: 83-93.

Evenson, G. R., Golden, H. E., Lane, C. R. et al., 2016. "An improved representation of geographically isolated wetlands in a watershed-scale hydrologic model." *Hydrol. Process* 30(22): 4168-4184

Flörke, M., Kynast, E., Bärlund, I., et al., 2013. "Domestic and industrial water uses of the past 60 years as a mirror of socio-economic development: A global simulation study." *Global. Enviro. Change* 23(1): 144-156.

Gao, J., Wang, R., Huang, J., 2015. "Ecological engineering for traditional Chinese agriculture-A case study of Beitang." *Ecol. Eng.* 76: 7-13.

Ghajarnia, N., Destouni, G., Thorslund, J., et al., 2020. "Data for wetlandscapes and their changes around the world." *Earth. Syst. Sci. Data* 12(2): 1083-1100.

Golden, H. E., Lane, C. R., Rajib, A., et al., 2021. "Improving global flood and drought predictions: Integrating non-floodplain wetlands into watershed hydrologic models." *Environ. Res. Lett.* 16(9):1-5.

Golden, H. E., Creed, I. F., Ali, G., et al., 2017. "Integrating geographically isolated wetlands into land management decisions." *Front. Ecol. Environ.* 15(6): 319-327.

Golden, H. E., Rajib, A., Lane, C. R., et al., 2019. "Non-floodplain wetlands affect watershed nutrient dynamics: A critical review." *Environ. Sci. Technol.* 53(13): 7203-7214.

Hill, M. J., Hassall, C., Oertli, B., et al., 2018. "New policy directions for global pond conservation." *Conserv. Lett.* 11(5): e12447.

Holgerson, M. A., and Raymond, P. A., 2016. "Large contribution to inland water CO2 and CH4 emissions from very small ponds." *Nat. Geosci.* 9(3): 222-226.

Hunter, J., Acuña, V., and Bauer, D. M., 2017. "Conserving small natural features with large ecological roles: A synthetic overview." *Biol. Conserv.* 211: 88-95.

Hu, S., Niu, Z., and Chen, Y., 2017. "Global wetland datasets: A review." *Wetlands* 37(5): 807-817.

Lane, C. R., Leibowitz, S. G., Autrey, B. C., et al., 2018. "Hydrological, physical, and chemical functions and connectivity of non-floodplain wetlands to downstream waters: A review." *J. Am. Water Resour. As.* 54(2): 346-371.

Lane, C. R., and D'Amico, E., 2016. "Identification of putative geographically isolated wetlands of the conterminous United States." *J. Am. Water Resour.*

As. 52(3): 705-722.

Lawford, R., Strauch, A., Toll, D., et al., 2013. "Earth observations for global water security." *Curr. Opin. Env. Sust.* 5(6): 633-643.

Lee, H., Jung, T. H., tom Dieck M. C., et al., 2020. "Experiencing immersive virtual reality in museums." *Inform. Manage.* 57(5): 103229.

Lee, L. H., Braud, T., Zhou, P., et al., 2021. "All one needs to know about metaverse: A complete survey on technological singularity, virtual ecosystem, and research agenda." arXiv: 2110.05352

Leibowitz, S. G., and Nadeau, T. L., 2003. "Isolated wetlands: State-of-the-science and future directions." *Wetlands.* 23: 663-684.

McDonald, C. P., Rover, J. A., Stets, E., G., et al. 2012. "The regional abundance and size distribution of lakes and reservoirs in the United States and implications for estimates of global lake extent." *Limnol. Oceanogr* 57(2): 597-606.

Ministry of Ecology and Environment of China, 2020. The 2020 annual statistical report of China's ecology and environment.

Mushet, D. M., Alexander, L. C., Bennett, M. et al., 2019. "Differing modes of biotic connectivity within freshwater ecosystem mosaics." *J. Am. Water Resour. As.* 55(2): 307-317.

Mushet, D. M., Calhoun, A. J., Alexander, L. C., et al., 2015, "Geographically isolated wetlands: Rethinking a misnomer." *Wetlands.* 35(3): 423-431.

Peel, M. C., Finlayson, B. L., and McMahon, T., 2007. "Updated world map of the Köppen-Geiger climate classification." *Hydrol. Earth Syst. Sc.* 11: 1633-1644.

Pekel, J. F., Cottam, A., Gorelick, N., et al., 2016. "High-resolution mapping of global surface water and its long-term changes." *Nature* 540: 418-422.

Poschlod, P., and Braun-Reichert, R., 2017. "Small natural features with large ecological roles in ancient agricultural landscapes of Central Europe-History, value, status, and conservation." *Biol. Conserv.* 211: 60-68.

Rains, M. C., Leibowitz, S. G., Cohen, M. J., et al., 2016. "Geographically isolated wetlands are part of the hydrological landscape." *Hydrol. Process*

30: 153-160.

Rajib, A., Golden, H. E., Lane, C. R., et al., 2020. "Surface depression and wetland water storage improves major river basin hydrologic predictions." *Water Resour. Res.* 56(7): e2019WR026561.

Salas, D., Liang, X., Navarro, M., et al., 2020. "An open-data open-model framework for hydrological models' integration, evaluation and application Environ." *Modell. Softw.* 126: 104622.

Sayer, C. D. and Greaves, H. M., 2020. "Making an impact on UK farmland pond conservation." *Aquat. Conserv.* 30(9): 1821-1828.

Stjepanović, S., Tomić, D. and Škare, M., 2017. "A new approach to measuring green GDP: A cross-country analysis." *Entrep. Sustain. Iss.* 4(4): 574-590.

Sullivan, S., Rains, M. C., Rodewald, A., et al., 2020. Distorting science, putting water at risk. Science, 369: 766-768.

Sullivan, S., Rains, M. C. and Rodewald, A., 2019. "Opinion: The proposed change to the definition of 'waters of the United States' flouts sound science." *P. Natl. Acad. Sci. USA.* 116(24): 11558-11561

Swartz, T. M., and Miller, J. R., 2021. "The American pond belt: An untold story of conservation challenges and opportunities." *Front. Ecol. Environ.* 19(9): 501-509.

Takeuchi, K., Ichikawa, K., and Elmqvist, T., 2016. "Satoyama landscape as social-ecological system: Historical changes and future perspective." *Curr. Opin. Env. Sust.* 19: 30-39.

Thorslund, J., Jarsjo, J., Jaramillo, F., et al., 2017. "Wetlands as large-scale nature-based solutions: Status and challenges for research, engineering and management." *Ecol. Eng.* 108: 489-497.

Wang, L., Tong, J., and Li, Y., 2019. "River chief system (RCS): An experiment on cross-sectoral coordination of watershed governance." *Fron. Env, Sci, Eng.* 13(4): 1-3.

Williams, R. T., Fryirs, K. A., and Hose, G. C., 2020. "The hydrological function of a large chain-of-ponds: A wetland system with intermittent surface flows." *Aquat. Sci.* 82(3): 1-18.

Wu, Q., Lane, C., R., Li, X., et al., 2019. "Integrating LiDAR data and multi-temporal aerial imagery to map wetland inundation dynamics using Google earth engine." *Remote Sens. Environ.* 228: 1-13.

Xu, W., Fan, X., Ma, J., et al., 2019. "Hidden loss of wetlands in China." *Curr. Biol.* 29(18): 3065e3071.

Yeo, I. Y., Lee, S., Lang, M. W., et al., 2019. "Mapping landscape-level hydrological connectivity of headwater wetlands to downstream waters: A catchment modeling approach-Part 2." *Sci. Total Environ.* 653: 1557-1570.

Zeng, L., and Chu, X., 2021. "Integrating depression storages and their spatial distribution in watershed-scale hydrologic modeling." *Adv. Water Resour.* 151: 103911.

Zhang, Z., Fluet-Chouinard, E., Jensen, K., et al., 2021. "Development of the global dataset of wetland area and dynamics for methane modeling (WAD2M)." *Earth Syst. Sci. Data* 13(5): 2001-2023.

Appendix

Section 9.1 Literature retrieval and wetland type analysis

We developed a dual-step search procedure to retrieve peer-reviewed articles on non-floodplain wetlands (NFWs; including main types and subtypes, as described below) from the Web of Science™ Core Collection. To capture the latest knowledge and scientific advances in this field, we focused on published literature within the most recent two decades; 2001-2020. Table A9.1 shows the search strings, supplementary instructions, and results of the search procedure, as detailed below.

Table A9.1 Dual-step search procedure for published articles on NFWs

	Search string and keyword	Supplementary instruction	Result
Step 1	non-floodplain wetland OR isolated wetland OR geographically isolated wetland OR small water bodies OR small vulnerable waters OR neglected freshwater habitat OR temporary wetland OR wetlandscape AND review OR commentary OR perspective	1) Each retrieved article was read through to ensure that it is a review, commentary, or perspective study of NFWs. 2) Studies on aquaculture and water treatment facilities were excluded. 3) Specific wetland types were manually screened and organized from the eligible articles.	57 articles and organized wetland types in Table A9.2
Step 2	each wetland type in Supplementary Table 2 AND wetland OR small water bodies OR small vulnerable waters OR neglected freshwater habitat OR wetlandscape	1) Only one keyword in the left parentheses and one keyword in the right parentheses were joined for each search. 2) Results were ranked by relevance. 3) Abstract, conclusions, and figures were perused for each retrieved article from top to bottom, to filter out articles that are not related to NFWs. 4) A maximum of 50 eligible articles were selected for each search.	2213 articles

Table A9.2 Main NFW types and subtypes, organized according to the inland wetland classification scheme of the Ramsar Convention and the U.S. Environmental Protection Agency.

Main type	Subtype	Main type	Subtype
Bogs	Kettle-holes	Ponds (continued)	Karst ponds
	Pocosins		Multi-pond systems
Constructed wetlands (CWs)			Temporary ponds
Cypress domes [a]		Pools	Temporary pools
Delmarva and Carolina bays [a]			Seasonal pools
Desert springs [a]			Vernal pools
Fens	Alvar wetlands	Potholes	Limesinks
	Wet meadows		Prairie potholes
Marshes			Sinkholes
Playas [a]	Playa lakes	Sandhills wetlands [a]	
Polar wetlands a		Seeps [a]	Desert seeps
Ponds	Chain-of-ponds		Hillside seeps
	Delmarva ponds	Swales [a]	
	Farm ponds	Swamps	

[a] These wetland types and their subtypes are organized into "other wetland types" in Fig. 9.1 in the main manuscript, due to the relatively smaller number of articles related to them (as described below).

In the first step, both "NFW" "isolated wetland" "geographically isolated wetland", and other keywords that reveal the characteristics and research focuses of these wetlands, were used to retrieve review, commentary, and perspective studies in the field. Studies on aquaculture and water treatment facilities were excluded because the ecosystem processes, particularly the hydroperiod of these waters, are largely human-controlled. Specific NFW types were then manually screened from the result 57 articles based on the authors' expertise as wetland scientists, and then organized into main wetland types and subtypes (Table A9.2). To make subsequent search steps concise and effective, common descriptive attributives (e.g., "blanket" in "blanket bogs" and "ephemeral" in "ephemeral pools") and equivalents (e.g., "constructed ponds or pools" that are equal to "constructed wetlands") were omitted here. Even so, the alphabetical collection of wetland types in Table A9.2 should be interpreted with caution — it is not a strict classification scheme for NFWs, as some terms of types are used as synonyms in the literature (e.g., "constructed wetlands" and "ponds", and "vernal pools" and "temporary pools").

In the second step, we expanded the search criteria by joining each wetland type in Table A9.2 and characteristic keywords used above, to retrieve all published articles on NFWs, i.e., not limited to review, commentary, and perspective studies. Some terms like "constructed wetlands" "marshes" and "ponds" had a large number of retrieved results (> 10,000), while the others like "Delmarva ponds" "multi-pond systems" and "limesinks" had, comparatively, very few results (< 30). To address this issue and evenly use the above wetland types, we ranked the retrieved articles by relevance according to the WoS built-in algorithm (September 30 2021 Release), and then reviewed the abstract, figures, and conclusions of each article manually, to ensure the remaining articles are related to NFWs, and filter out articles that are not related to NFWs. In other words, with particular attention paid to geographical features (i.e., non-floodplain), articles were retained only when the research object belongs to, contains, is within NFWs, or the research topic is explicitly related to NFWs. Otherwise, a minimum distance of 10-meters far away from rivers, coasts, and floodplains was used as a criterion for the visual inspection of NFWs, according to Lane and D'Amico, 2016. Based on this procedure and filtering out repetitive articles that have already been retrieved, the top 50 eligible articles (or all articles if less than 50) were selected for each search, resulting in a total dataset of 2,213 articles for this step. The maximum number of 50 was used to balance the number and relevance of articles for each retrieval.

Based on the full set of literature, we analyzed the number of articles where each NFW type (main and subtypes) occurs and the proportion of studies (main types) on each continent (Fig. 9.1 in the main manuscript). The number of articles was counted by using the advance search tool of Zotero. Results were shown in a grouped tag cloud (Fig. 9.1 in the main manuscript), rather than standard statistical charts (e.g., pie charts), because, as mentioned above, the wetland types we organized here were not a strict classification scheme for NFWs, while some articles were double-counted in the analysis. Specifically, when calculating the proportions of wetland types being studied, the number of continent-related articles increased by one, if a review study includes a discussion of this continent, or a case study is conducted on this continent. This section is a meta-analysis and not a systematic review. The statistical results should be interpreted with caution, due to the threshold in selecting the 50 most relevant articles for each search of each wetland type. Future work would potentially capture more comprehensive studies via a clearly elaborated NFW classification scheme, as proposed in the section of new opportunities for the sustainable NFW use in the main manuscript.

Section 9.2 Representative region selection and land-cover analysis

In addition to the meta-analysis, 30 representative regions were selected and grouped to investigate research characteristics across NFWs around the world. Following the selection criteria narrated in the main manuscript, the active projects around NFW were gathered from information within the European Pond Conservation Network, Global Wetland Ecohydrology Network, and National Association of Wetland Managers. Both classification and clustering experiments were conducted to group the representative NFW regions when investigating their land-cover types and research characteristics. The classification was based on the assumption that NFWs play different roles across different continents, land-cover characteristics, or hydro-climatic settings (by the updated Köppen-Geiger climate classification system), while the clustering, as a type of unsupervised learning, was performed by the Hierarchical Cluster Analysis tools of SPSS Statistics 26. These regions were eventually grouped by the climate zones due to the most explanatory results and widely-recognized hydro-climatic drivers of wetland change at global scales (Bertassello et al., 2019; Ahlén et al., 2021).

Satellite imagery (Fig. A9.1; showing NFWs and surrounding landscapes), climate group, wetland type, and location information (Supplementary Table 3) of the 30 representative regions are presented, with the same identifiers (IDs) as in Fig. 9.2 in the main manuscript and in Table A9.3, A9.4 and A9.7 below. Additionally, on-site photos of similar NFWs and surrounding landscapes can be found in Google Images by searching for the specific wetland types and locations in Table A9.3. They can serve as a reference for the general readership. The grid-formatted Esri 2020 Global Land Cover map was used to analyze the proportion of land-cover types for each site. This dataset was built on the European Space Agency Sentinel-2 satellite imagery and has a recent release date (July 2021) and high spatial resolution (10-meter) at global scales. The land-cover map was clipped by twice the extent of the imagery for each site in Fig. A9.1 to better grasp the land-cover characteristics around NFWs, and then reclassified and removed null values (i.e., invalid values in the source files) using ArcToolbox, yielding the proportional results in Table A9.4. Due to the relatively small sample size in each group, an independent-samples t-test was performed using SPSS, to determine whether the proportions of one climate zone were statistically higher (p-value < 0.1) than those of the other climate zones in the same land-use type.

Fig. A9.1 Recent (July 2021) satellite imagery of the 30 representative NFW regions (numbered according to Site ID). Data source: Google Earth and Maxar Technologies.

Table A9.3 Information about the 30 representative NFW regions, including their distribution across the climate zones used, region name with specific wetland type, and detailed location (including coordinates).

Climate zone	Region ID	Region name	Detailed location	Central coordinates of the satellite imagery
Tropical	1	Mangrove swamps-North Colombia [a]	Near the Ciénaga Grande de Santa Marta, Colombia	74.58W, 10.49N
	2	Swamps-East Brazil	Rural area of Canindé, state of Ceará, Brazil	39.27W, 4.35S
	3	Swamps-Congo	Sangha river basin, Congo	17.31E, 0.45N
	4	Temporary pools-Tanzania	Kilombero river valley, Tanzania	36.05E, 8.87S
	5	Marshes-Sri Lanka	Jaffna peninsula, Sri Lanka	79.99E, 9.74N
	6	Swamps-Indonesia	Sentarum Lake National Park, Indonesia	112.14E, 0.99N
Arid	7	Vernal pools-West U.S.	Sierra Nevada mountain range of California, U.S.	118.81W, 36.21N
	8	Swamps-North Mexico	Northeast state of Coahuila, Mexico	100.90W, 28.99N
	9	Playa lakes-Spain	Northwest Malaga Province, Spain	4.85W, 37.09N

Continued

Climate zone	Region ID	Region name	Detailed location	Central coordinates of the satellite imagery
Arid	10	Swamps-Libya	South of the Maradah oasis, Libya	19.74E, 28.93N
Arid	11	Temporary pools-Uzbekistan	Amu Darya river delta, Uzbekistan	58.87E, 42.94N
Arid	12	Playas-Central Australia	Between the Lake White and Gregory, Australia	128.42E, 20.87S
Temperate	13	Mangrove swamps-Southeast U.S. [a]	Peace River Basin, central Florida, U.S.	81.67W, 28.44N
Temperate	14	Temporary ponds-England	Northeast Norfolk, England	1.16E, 52.91N
Temperate	15	Temporary ponds-South Africa	Southeast suburb of Cape Town, South Africa	18.56E, 34.03S
Temperate	16	Marshes-North India	Jaunpur rural district, Uttar Pradesh, India	82.90E, 25.92N
Temperate	17	Farm ponds-South China	Xifu River Watershed, Guangdong Province, China	114.76E, 23.05N
Temperate	18	Multi-pond systems-Southeast China	Huashan Watershed, Anhui Province, China	118.20E, 32.32N
Temperate	19	Farm ponds-Japan	North of Himeji City, Honshu Japan	134.99E, 34.94N
Temperate	20	Chain-of-ponds-Southeast Australia	Southern Tablelands of New South Wales, Australia	149.63E, 34.87S
Cold	21	Ponds-Alaska	West of Lake Iliamna, Alaska, U.S.	156.23W, 59.32N
Cold	22	Prairie potholes-Canada	Southern Saskatchewan, Canada	106.55W, 50.74N
Cold	23	Peat bogs-Canada	Hudson Bay Lowlands, Ontario, Canada	85.19W, 53.29N
Cold	24	Marshes-Sweden	Norrström drainage basin, Sweden	14.29E, 60.23N
Cold	25	Peat bogs-West Russia	Near the Volga River Delta, Russia	46.29E, 44.44N
Cold	26	Peat bogs-Siberia	Lena river basin, Siberia (Russian Far East)	121.71E, 66.59N

Climate zone	Region ID	Region name	Detailed location	Central coordinates of the satellite imagery
Polar	27	Polar wetlands-West Antarctica	King George Island, Antarctica	58.48W, 62.17S
Polar	28	Polar wetlands-Greenland	Disko Island, west central Greenland	52.15W, 69.54N
Polar	29	Ponds-Qinghai-Tibet Plateau	East of Lake Namuka Co, Qinghai-Tibet Plateau (the Third Pole)	90.16E, 31.92N
Polar	30	Polar wetlands-East Antarctica	The McMurdo Dry Valley Region, Antarctica b	163.32E, 78.28S

[a] Note that in addition to coastal areas, mangroves exist in some inland, historic and isolated wetland systems, according to Jaramillo et al. (2018) and Aburto-Oropeza et al. (2021).

[b] Upland ponds fed by intermittent, meltwater streams exist in the McMurdo Dry Valleys, according to Jungblut et al. (2012) and Wlostowski et al. (2016).

Table A9.4 Classified land-cover proportions (%) around the 30 representative NFW regions, quantified by using the grid-formatted Esri 2020 Global Land Cover map.

Climate zone	Site ID	Water	Trees	Grass	Flooded vegetation	Crops	Scrub / Shrub	Built area	Bare ground	Snow / Ice
Tropical	1	2.19	5.22	14.16	36.22	1.48	40.37	0.36	0.00	0.00
	2	0.73	0.56	0.02	0.00	0.02	95.26	3.41	0.00	0.00
	3	1.10	61.82	25.96	7.76	0.00	3.28	0.08	0.00	0.00
	4	1.81	0.29	1.80	15.43	46.57	28.88	5.23	0.00	0.00
	5	1.34	0.26	0.31	0.26	29.13	0.33	68.36	0.00	0.00
	6	7.02	47.29	0.00	43.60	0.00	2.09	0.00	0.00	0.00
	Average	2.37	19.24	7.04	17.21	12.87	28.37	12.91	0.00	0.00
Arid	7	0.09	5.75	0.19	0.00	0.00	90.97 ** [a]	2.89	0.10	0.10
	8	1.00	2.28	0.00	0.00	9.87	83.44 **	2.99	0.41	0.00
	9	1.12	3.22	0.07	0.00	37.58	52.26 **	3.32	2.43	0.00
	10	0.22	0.00	0.00	0.00	0.00	1.02 **	0.02	98.75	0.00
	11	0.42	0.00	0.00	0.00	74.01	17.77 **	7.54	0.26	0.00
	12	0.00	3.02	0.00	0.00	0.00	96.93 **	0.00	0.04	0.00
	Average	0.47	2.38	0.04	0.00	20.24	57.06	2.79	17.00	0.02

Continued

Climate zone	Site ID	Water	Trees	Grass	Flooded vegetation	Crops	Scrub / Shrub	Built area	Bare ground	Snow / Ice
Temperate	13	12.64	30.13 *	7.59	1.71	1.95 **	14.11	31.02 *	0.85	0.00
	14	0.07	20.67 *	0.15	0.00	74.94 **	0.58	3.59 *	0.00	0.00
	15	0.21	4.54 *	2.44	0.12	31.24 **	4.72	56.33 *	0.39	0.00
	16	0.84	0.25 *	0.04	0.16	80.32 **	0.58	17.81 *	0.00	0.00
	17	15.00	59.52 *	0.19	0.00	18.69 **	0.54	6.05 *	0.00	0.00
	18	4.63	28.10 *	0.00	0.00	55.74 **	0.11	11.42 *	0.00	0.00
	19	3.24	45.16 *	2.16	0.00	11.03 **	0.85	37.56 *	0.00	0.00
	20	0.06	2.79 *	60.46	0.00	27.18 **	8.27	1.24 *	0.00	0.00
	Average	4.59	23.89	9.13	0.25	37.64	3.72	20.63	0.16	0.00
Cold	21	5.64 *	7.38	13.83	49.58	0.00	23.57	0.00	0.00	0.00
	22	0.31 *	0.02	0.00	5.81	89.81	3.65	0.39	0.00	0.00
	23	8.86 *	1.40	1.31	84.21	0.00	4.19	0.00	0.04	0.00
	24	8.27 *	76.17	1.86	0.23	0.00	13.13	0.34	0.00	0.00
	25	0.48 *	0.00	9.16	0.00	0.94	84.14	0.00	0.00	5.28
	26	11.89 *	33.75	3.75	5.08	0.06	0.06	45.28	0.20	0.00
	Average	5.91	19.79	4.98	24.15	15.14	21.46	7.67	0.04	0.88
Polar	27	2.74	0.91	0.09	0.00	0.00	19.57	0.07	34.60 **	42.03 *
	28	0.30	0.76	0.61	0.90	0.00	91.56	0.00	4.72 **	1.15 *
	29	2.99	0.00	0.00	0.08	0.00	70.00	0.00	26.92 **	0.00 *
	30	0.00	0.00	0.00	0.00	0.00	0.00	0.00	38.78 **	61.22 *
	Average	1.51	0.42	0.18	0.25	0.00	45.28	0.02	26.25	26.10

[a] The one and two asterisks indicate statistical significance with confidence intervals of 90 and 95%, respectively.

Section 9.3 Investigation of research focuses

Twelve investigated research focuses were summarized according to the full set of literature retrieved above, the Millennium Ecosystem Assessment (WHO, 2006), and the authors' expertise in wetland sciences (Table A9.5). These research focuses were used to reflect the main scientific and practical concerns on the sustainable NFW use, despite a few overlaps, e.g., among the hydrological and biogeochemical patterns and processes. Meanwhile, for each of the 30 investigated regions, we considered additional site-related articles to quantify the prevalence of the 12 classes of research focuses (Table A9.6). These articles were selected from the full set of literature, based on the criteria covering the representative regions in Fig. A9.1 and surrounding areas with the same type of wetland. Based on a thorough reading, the abstract and text of each article, excluding references, were used for manual judgment. Table A9.7 shows the aggregated proportions of research focus classes for the 30 regions, categorized according to the five climate zones. The proportional values were organized into three groups by size using the Jenks Natural Breaks Classification method implemented in ArcGIS, to conduct the analysis of research focuses in the main manuscript.

Table A9.5 Investigated research focuses of NFWs and surrounding landscapes

Research focus	Explanations
Hydrogeomorphology (HM)	Hydrological regime and geomorphic settings that shape and develop the various wetland types and further influence their functions
Water resource (WR)	Natural resources of water in wetlands (e.g., storage capacity) that are useful or potentially useful to agricultural, industrial, domestic and other human activities
Hydrological processes (HP)	Components that influence the water budget, including precipitation, evaporation, surface and subsurface inflows, spillage and seepage outflows, groundwater recharge and discharge, etc.

Continued

Research focus	Explanations
Biogeochemical processes (BP)	Complex interactions between HP, mineralogical transformations, bacterial and vegetation communities, and soil stores of carbon and nutrients, including the regulation/cycling of carbon, nitrogen, phosphorus, sulfur, mercury, and other elements
Nutrient retention (NR)	Role of wetlands as a sink that reduce nutrient loads for downstream waters. This item is valid only when the term "retention", "reduction", or "removal" is mentioned, as relevant mechanisms, including sedimentation, plant uptake, and microbial decomposition, were classified to belong to BP
Greenhouse gas regulation (GG)	Fluxes of greenhouses gases (e.g., carbon dioxide, methane, and nitrous oxide), emitted into and absorbed from the atmosphere, and the driving mechanisms like photosynthesis, respiration, decomposition, nitrification, and denitrification
Biodiversity (BD)	The large number and variety of plants, animals, and microorganisms that wetlands support, as well as the food, habitats, and breeding grounds for these creatures that wetlands provide
Ecosystem services (ES)	The four categories of provisioning, regulating, cultural and supporting services, defined by the Millennium Ecosystem Assessment (WHO, 2006). This item is valid only when the term "ecosystem services" is mentioned, i.e., these services are recognized as a whole
Sustainable agriculture (SA)	Wetland-related agriculture that holds a multi-pronged goal: a healthy environment, sufficient food production, economic profitability, and good quality of life for the practitioners. This item is valid when any of the goals is mentioned
Inventory mapping (IM)	Dataset and digitized map that provides information on the actual or potential location, size, and type of wetlands
Climate change (CC)	Global warming caused by human activity, particularly the burning of fossil fuels and removal of forests, and its impacts on Earth's weather patterns. This item is valid only when CC is associated with NFWs in the literature
Wetland protection (WP)	Protecting and conserving the areas where the wetland exist. This item is valid only when the term "protection", "conservation", or "preservation" is mentioned

Table A9.6 Site-related literature, including their research focuses, of the 30 NFW representative regions.

Site ID	Id	Article	HM	WR	HP	BP	NR	GG	BD	ES	SA	IM	CC	WC
1	1	Craven, J. et al. Development and testing of a river basin management simulation game for integrated management of the Magdalena-Cauca river basin. Environ. Modell. Softw. 90, 78-88 (2017).		√ [a]	√					√			√	√
	2	Garcés, O. O. et al. Marine litter and microplastic pollution on mangrove soils of the Ciénaga Grande de Santa Marta, Colombian Caribbean. Mar. Pollut. Bull. 145, 455-462 (2019).				√			√					√
	3	Jaramillo, F. et al. Effects of hydroclimatic change and rehabilitation activities on salinity and mangroves in the Ciénaga Grande de Santa Marta, Colombia. Wetlands 38, 755-767 (2018).			√		√			√			√	√
	4	Konnerup, D. et al. Nitrous oxide and methane emissions from the restored mangrove ecosystem of the Ciénaga Grande de Santa Marta, Colombia. Estuar. Coast. Shelf S. 140, 43-51 (2014).				√		√						√
	5	Polanía, J. et al. Recent advances in understanding Colombian mangroves. Acta. Oecol. 63, 82-90 (2015).	√						√				√	√
	6	Rivera, M. et al. Salinity and chlorophyll a as performance measures to rehabilitate a mangrove-dominated deltaic coastal region: The Ciénaga Grande de Santa Marta-Pajarales Lagoon Complex, Colombia. Estuar. Coast. 34, 1-19 (2011).			√	√								√
	7	Wemple, B. C. et al. Ecohydrological disturbances associated with roads: Current knowledge, research needs, and management concerns with reference to the tropics. Ecohydrology 11, e1881 (2018).			√				√				√	√
	8	Zamora, S. et al. Carbon fluxes and stocks by Mexican tropical forested wetland soils: A critical review of its role for climate change Mitigation. Int. J. Env. Res. Pub. He. 17, 7372 (2020).						√		√			√	√
	9	Zipper, S. C. et al. Integrating the water planetary boundary with water management from local to global scales. Earths Future 8, e2019EF001377 (2020).	√	√	√		√	√	√	√			√	
2	1	Cunha, D. et al. Landscape patterns influence nutrient concentrations in aquatic systems: Citizen science data from Brazil and Mexico. Freshw. Sci. 38, 365-378 (2019).				√								
	2	Rodrigues, L. N. et al. Estimation of small reservoir storage capacities with remote sensing in the Brazilian savannah region. Water Resour. Manag. 26, 873-882 (2012).		√										
	3	Tabosa, A. et al. Live fast and die young: The aquatic macrophyte dynamics in a temporary pool in the Brazilian semiarid region. Aquat. Bot. 102, 71-78 (2012).							√					
	4	Tavares, C. et al. Environmental drivers of tadpole community structure in temporary and permanent ponds. Limnologica 81, 125764 (2020).							√					√
3	1	Alsdorf, D. et al. Opportunities for hydrologic research in the Congo basin. Rev. Geophys. 54, 378-409 (2016).	√	√	√							√	√	
	2	Bwangoy, J. R. B. et al. Wetland mapping in the Congo Basin using optical and radar remotely sensed data and derived topographical indices. Remote. Sens. Environ. 114, 73-86 (2010).			√			√	√	√		√		√

Continued

Site ID	Id	Article	HM	WR	HP	BP	NR	GG	BD	ES	SA	IM	CC	WC
3	3	Bwangoy, J. R. B. et al. Identifying nascent wetland forest conversion in the democratic republic of the Congo. Wetl. Ecol. Manag. 21, 29-43 (2013).							√	√		√		√
	4	Cotterill, F. P. D. The Upemba lechwe, Kobus anselli: An antelope new to science emphasizes the conservation importance of Katanga, democratic republic of Congo. J. Zool. 265, 113-132 (2005).	√						√	√				√
	5	Dinerstein, E. et al. A global deal for nature: Guiding principles, milestones, and targets. Sci. Adv. 5, eaaw2869 (2019).						√					√	
	6	Grainger, A. and Kim, J. Reducing global environmental uncertainties in reports of tropical forest carbon fluxes to REDD+ and the Paris agreement global stocktake. Remote. Sens. 12, 2369 (2020).						√					√	
	7	Jung, H. C. et al. Characterization of complex fluvial systems using remote sensing of spatial and temporal water level variations in the Amazon, Congo, and Brahmaputra rivers. Earth. Surf. Proc. Land. 35, 294-304 (2010).	√	√	√									
	8	Keddy, P. A. et al. Wet and wonderful: The world's largest wetlands are conservation priorities. Bioscience 59, 39-51 (2009).			√	√	√	√	√	√		√	√	√
	9	Mitchard, E. T. The tropical forest carbon cycle and climate change. Nature 559, 527-534 (2018).				√		√	√	√		√	√	√
4	1	Lyon, S. W. et al. Interpreting characteristic drainage timescale variability across Kilombero valley, Tanzania. Hydrol. Process. 29, 1912-1924 (2015).		√	√						√			
	2	Koutsouris, A. J. et al. Comparing global precipitation data sets in eastern Africa: A case study of Kilombero valley, Tanzania. Int. J. Climatol. 36, 2000-2014 (2016).		√	√									
5	1	Gopalakrishnan, T. et al. Sustainability of coastal agriculture under climate change. Sustainability 11, 7200 (2019).						√	√				√	
	2	Halwatura, D. and Najim, M. Application of the HEC-HMS model for runoff simulation in a tropical catchment. Environ. Modell. Softw. 46, 155-162 (2013).			√									
	3	Ramasamy, R. and Surendran, S. N. Global climate change and its potential impact on disease transmission by salinity-tolerant mosquito vectors in coastal zones. Front. Physiol. 3, 198 (2012).						√	√				√	
	4	Rebelo, L. M. et al. Remote sensing and GIS for wetland inventory, mapping and change analysis. J. Environ. Manage. 90, 2144-2153 (2009).								√		√		√
	5	Santiapillai, C. and de Silva, M. Status, distribution and conservation of crocodiles in Sri Lanka. Biol. Conserv. 97, 305-318 (2001).							√					
6	1	Chen, G. C. et al. Rich soil carbon and nitrogen but low atmospheric greenhouse gas fluxes from North Sulawesi mangrove swamps in Indonesia. Sci. Total. Environ. 487, 91-96 (2014).					√		√					
	2	Couwenberg, J. et al. Greenhouse gas fluxes from tropical peatlands in south-east Asia. Global. Change Biol. 16, 1715-1732 (2010).						√					√	
	3	Dohong, A. et al. A review of the drivers of tropical peatland degradation in south-east Asia. Land use policy 69, 349-360 (2017).			√			√	√	√			√	√
	4	Graham, L. L. et al. A common-sense approach to tropical peat swamp forest restoration in Southeast Asia. Restor. Ecol. 25, 312-321 (2017).			√	√		√	√	√				
	5	Hergoualc'h, K., and Verchot, L. V. Stocks and fluxes of carbon associated with land use change in south-east Asian tropical peatlands: A review. Global. Biogeochem. Cy. 25, (2011).					√		√			√	√	

296 ● Hydro-Environmental Assessment of Small Water Bodies:
From Local to Global Scales

Site ID	Id	Article	HM	WR	HP	BP	NR	GG	BD	ES	SA	IM	CC	WC
6	6	Kumar, P. et al. Towards an improved understanding of greenhouse gas emissions and fluxes in tropical peatlands of Southeast Asia. Sustain. Cities. Soc. 53, 101881 (2020).	√		√	√		√	√	√	√		√	√
	7	Margono, B. A. et al. Mapping wetlands in Indonesia using Landsat and PALSAR data-sets and derived topographical indices. Geo-spat. Inf. Sci. 17, 60-71 (2014).						√	√	√		√		
	8	Murdiyarso, D. et al. Opportunities for reducing greenhouse gas emissions in tropical peatlands. P. Natl. Acad. Sci. 107, 19655-19660 (2010).						√	√				√	√
	9	Page, S. E. et al. Global and regional importance of the tropical peatland carbon pool. Global. Change Biol. 17, 798-818 (2011).						√				√	√	
	10	Warren, M. et al. An appraisal of Indonesia's immense peat carbon stock using national peatland maps: Uncertainties and potential losses from conversion. Carbon. Bal. Manage. 12, 1-12 (2017).						√	√	√			√	√
7	1	Black, C. H. et al. Using wildfires as a natural experiment to evaluate the effect of fire on southern California vernal pool plant communities. Glob. Ecol. Conserv. 7, 97-106 (2016).							√	√				√
	2	Gosejohan, M. C. et al. Hydrologic influences on plant community structure in vernal pools of northeastern California. Wetlands 37, 257-268 (2017).			√			√						√
	3	Huntsinger, L. and Oviedo, J. L. Ecosystem services are social-ecological services in a traditional pastoral system: The case of California's Mediterranean rangelands. Ecol. Soc. 19, 8 (2014).				√		√	√	√				√
	4	Kneitel, J. M. et al. California vernal pool endemic responses to hydroperiod, plant thatch, and nutrients. Hydrobiologia 801, 129-140 (2017).			√	√		√	√					√
	5	Merriam, K. E. et al. Livestock use has mixed effects on slender orcutt grass in Northeastern California Vernal Pools. Rangeland. Ecol. Manag. 69, 185-194 (2016).							√		√			√
	6	Pyke, C. R. Assessing climate change impacts on vernal pool ecosystems and endemic branchiopods. Ecosystems 8, 95-105 (2005).				√							√	√
	7	Raimondo, S. et al. A unified approach for protecting listed species and ecosystem services in isolated wetlands using community-level protection goals. Sci. Total. Environ. 663, 465-478 (2019).			√	√		√	√					√
	8	Rains, M. C. et al. Geological control of physical and chemical hydrology in California vernal pools. Wetlands 28, 347-362 (2008).		√	√	√		√						√
	9	Rains, M. C. et al. The role of perched aquifers in hydrological connectivity and biogeochemical processes in vernal pool landscapes, Central Valley, California. Hydrol. Process. 20, 1157-1175 (2006).			√	√								√
	10	Rice, K. J. and Emery, N. C. Managing microevolution: Restoration in the face of global change. Front. Ecol. Environ. 1, 469-478 (2003).								√			√	√
	11	Sinnathamby, S. et al. A sensitivity analysis of pesticide concentrations in California Central Valley vernal pools. Environ. Pollut. 257, 113486 (2020).				√						√		
	12	Sloop, C. M. et al. Conservation genetics of butte county meadowfoam (Limnanthes floccose ssp. Californica Arroyo), an endangered vernal pool endemic. Conserv. Genet. 12, 311-323 (2011).								√				√

Continued

Site ID	Id	Article	HM	WR	HP	BP	NR	GG	BD	ES	SA	IM	CC	WC
		Site-related literature			**Research focus**										
7	13	Varin, M. et al. Mapping vernal pools using lidar data and multitemporal satellite imagery. Wetlands 41, 1-15 (2021).			√							√		√
	14	Wacker, M. and Kelly, N. M. Changes in vernal pool edaphic settings through mitigation at the project and landscape scale. Wetl. Ecol. Manag. 12, 165-178 (2004).	√							√				√
8	1	Beach, T. et al. A review of human and natural changes in Maya lowland wetlands over the Holocene. Quaternary. Sci. Rev. 28, 1710-1724 (2009).	√							√	√			
	2	Howeth, J. G. et al. Contrasting demographic and genetic estimates of dispersal in the endangered Coahuilan box turtle: a contemporary approach to conservation. Mol. Ecol. 17, 4209-4221 (2008).							√					
	3	Krause, S. et al. Ancient Maya wetland management in two watersheds in Belize: Soils, water, and paleoenvironmental change. Quatern. Int. 502, 280-295 (2019).	√			√				√				
	4	Minckley, T. A. et al. The relevance of wetland conservation in arid regions: A re-examination of vanishing communities in the American Southwest. J. Arid. Environ. 88, 213-221 (2013).		√					√	√	√		√	√
9	1	Höbig, N. et al. Palaeohydrological evolution and implications for palaeoclimate since the Late Glacial at Laguna de Fuente de Piedra, southern Spain. Quatern. Int. 407, 29-46 (2016).	√			√								
	2	Kohfahl, C. et al. Characterising flow regime and interrelation between surface-water and ground-water in the Fuente de Piedra salt lake basin by means of stable isotopes, hydrogeochemical and hydraulic data. J. Hydrol. 351, 170-187 (2008).		√	√	√								
	3	Montalván, F. J. et al. Hydrochemical and isotopes studies in a hypersaline wetland to define the hydrogeological conceptual model: Fuente de Piedra Lake (Malaga, Spain). Sci. Total Environ. 576, 335-346 (2017).				√								
	4	Rodríguez, R. M. et al. Applying piezometric evolution indicators to facilitate stakeholder's participation in the management of groundwater-dependent ecosystems. Case study: Fuente de Piedra playa lake (southern Spain). Hydrobiologia 782, 145-154 (2016).		√	√					√				
	5	Rodríguez, R. M. et al. Estimation of ground-water exchange with semi-arid playa lakes (Antequera region, southern Spain). J. Arid Environ. 66, 272-289 (2006).	√	√	√	√				√				
	6	Rodríguez, R. M. et al. Hydrogeological behaviour of the Fuente-de-Piedra playa lake and tectonic origin of its basin (Malaga, southern Spain). J. Hydrol. 543, 462-476 (2016).	√		√	√					√			
	7	Rodríguez, R. M. Hydrogeology of ponds, pools, and playa-lakes of southern Spain. Wetlands 27, 819-830 (2007).	√		√	√								√
	8	Park, L. E., and Gierlowski, E. H. Paleozoic lake faunas: Establishing aquatic life on land. Palaeogeogr. Palaeocl. 249, 160-179 (2007).			√			√	√			√		
	9	Sánchez, C. S. et al. Evapotranspiration in semi-arid wetlands: Relationships between inundation and the macrophyte-cover: Open-water ratio. Adv. Water Resour. 27, 643-655 (2004).			√									
10	1	Argyriou, T. et al. A fish assemblage from an early Miocene horizon from Jabal Zaltan, Libya. J. Afr. Earth. Sci. 102, 86-101 (2015).							√					

Site ID	Id	Article	HM	WR	HP	BP	NR	GG	BD	ES	SA	IM	CC	WC
10	2	Merken, R. et al. Wetland suitability and connectivity for trans-Saharan migratory waterbirds. PLoS One 10, p.e0135445 (2015).			√				√	√		√	√	√
11	1	Benduhn, F. and Renard, P., A dynamic model of the Aral Sea water and salt balance. J. Marine. Syst. 47, 35-50 (2004).			√									
	2	Jarsjö, J. et al. Climate-driven change of nitrogen retention-attenuation near irrigated fields: Multi-model projections for Central Asia. Environ. Earth Sci. 76, 1-12 (2017).		√	√	√	√				√		√	
	3	Törnqvist, R. et al. Water savings through improved irrigation techniques: basin-scale quantification in semi-arid environments. Water Resour. Manag. 26, 949-962 (2012).		√	√						√			
	4	Törnqvist, R. et al. Mechanisms of basin-scale nitrogen load reductions under intensified irrigated agriculture. PLoS One 10, e0120015 (2015).					√				√			
12	1	Box, J. B. et al. Central Australian waterbodies: The importance of permanence in a desert landscape. J. Arid. Environ. 72, 1395-1413 (2008).	√	√	√				√				√	√
	2	Cohen, T. J. et al. Continental aridification and the vanishing of Australia's megalakes. Geology 39, 167-170 (2011).			√									
	3	Gouramanis, C. et al. High-resolution, multiproxy palaeoenvironmental changes recorded from two mile lake, southern western Australia: Implications for Ramsar-listed playa sites. Mar. Freshwater. Res. 67, 748-770 (2015).	√		√	√			√					√
	4	Habeck, A. and Nanson, G. C. Environmental character and history of the Lake Eyre Basin, one seventh of the Australian continent. Earth-Sci. Rev. 132, 39-66 (2014).			√									
13	1	Batzer, D. P. The seemingly intractable ecological responses of invertebrates in north American wetlands: A review. Wetlands 33, 1-15 (2013).				√			√					√
	2	Chadwick, J. Integrated LiDAR and IKONOS multispectral imagery for mapping mangrove distribution and physical properties. Int. J. Remote. Sens. 32, 6765-6781 (2011).										√	√	
	3	Engle, V. D. Estimating the provision of ecosystem services by Gulf of Mexico coastal wetlands. Wetlands 31, 179-193 (2011).				√	√	√		√			√	√
	4	Lane, C. R. and D'Amico, E. Calculating the ecosystem service of water storage in isolated wetlands using LiDAR in north central Florida, USA. Wetlands 30, 967-977 (2010).	√	√	√					√				
	5	Lane, C. R. et al. Denitrification potential in geographically isolated wetlands of North Carolina and Florida, USA. Wetlands 35, 459-471 (2015).					√	√	√			√		
	6	McLaughlin, D. L. and Cohen, M. J. Ecosystem specific yield for estimating evapotranspiration and groundwater exchange from diel surface water variation. Hydrol. Process. 28, 1495-1506 (2014).			√					√				
	7	Middleton, B. A. Regeneration potential of Taxodium distichum swamps and climate change. Plant. Ecol. 202, 257-274 (2009).			√				√				√	
	8	Park, J. et al. Stochastic modeling of hydrologic variability of geographically isolated wetlands: Effects of hydro-climatic forcing and wetland bathymetry. Adv. Water Resour. 69, 38-48 (2014).	√	√	√					√				
	9	Reif, M. et al. Mapping isolated wetlands in a karst landscape: GIS and remote sensing methods. Gisci. Remote. Sens. 46, 187-211 (2009).	√									√		

Continued

<table>
<tr><th rowspan="2">Site ID</th><th colspan="2">Site-related literature</th><th colspan="12">Research focus</th></tr>
<tr><th>Id</th><th>Article</th><th>HM</th><th>WR</th><th>HP</th><th>BP</th><th>NR</th><th>GG</th><th>BD</th><th>ES</th><th>SA</th><th>IM</th><th>CC</th><th>WC</th></tr>
<tr><td rowspan="2">13</td><td>10</td><td>Said, A. et al. Simulation of surface water for un-gauged areas with storage-attenuation wetlands. J. Am. Water Resour. As. 43, 546-556 (2007).</td><td>√</td><td>√</td><td></td><td></td><td></td><td></td><td></td><td></td><td>√</td><td></td><td></td><td></td></tr>
<tr><td>11</td><td>Zhang, J. et al. Calibration of the HSPF model with a new coupled FTABLE generation method. Prog. Nat. Sci. 19, 1747-1755 (2009).</td><td></td><td></td><td>√</td><td></td><td></td><td></td><td></td><td></td><td></td><td></td><td></td><td></td></tr>
<tr><td rowspan="10">14</td><td>1</td><td>Biggs, J. et al. 15 years of pond assessment in Britain: Results and lessons learned from the work of pond conservation. Aquat. Conserv. 15, 693-714 (2005).</td><td></td><td></td><td></td><td></td><td>√</td><td></td><td>√</td><td></td><td></td><td></td><td></td><td>√</td></tr>
<tr><td>2</td><td>Bilton, D. T. et al. Ecology and conservation status of temporary and fluctuating ponds in two areas of southern England. Aquat. Conserv. 19, 134-146 (2009).</td><td></td><td></td><td></td><td></td><td>√</td><td></td><td>√</td><td></td><td></td><td></td><td></td><td>√</td></tr>
<tr><td>3</td><td>Ewald, N. C. et al. Climate change and trophic interactions in model temporary pond systems: The effects of high temperature on predation rate depend on prey size and density. Freshwater. Biol. 58, 2481-2493 (2013).</td><td></td><td></td><td></td><td></td><td></td><td></td><td>√</td><td></td><td></td><td></td><td>√</td><td>√</td></tr>
<tr><td>4</td><td>Lewis, J. et al. Pond management enhances the local abundance and species richness of farmland bird communities. Agr. Ecosyst. Environ. 273, 130-140 (2019).</td><td></td><td></td><td></td><td></td><td></td><td></td><td>√</td><td>√</td><td></td><td></td><td></td><td>√</td></tr>
<tr><td>5</td><td>Ruse, L. P., Greaves et al. Consequences of pond management for chironomid assemblages and diversity in English farmland ponds. J. Limnol. 77 (2018).</td><td></td><td></td><td></td><td></td><td></td><td></td><td>√</td><td></td><td></td><td></td><td></td><td>√</td></tr>
<tr><td>6</td><td>Sayer, C. and Greaves, H. Making an impact on UK farmland pond conservation. Aquat. Conserv. 30, 1821-1828 (2020).</td><td></td><td></td><td></td><td></td><td></td><td></td><td>√</td><td>√</td><td></td><td></td><td></td><td>√</td></tr>
<tr><td>7</td><td>Sayer, C. Conservation of aquatic landscapes: Ponds, lakes, and rivers as integrated systems. Wires. Water 1, 573-585 (2014).</td><td></td><td></td><td></td><td></td><td>√</td><td></td><td>√</td><td>√</td><td></td><td></td><td></td><td>√</td></tr>
<tr><td>8</td><td>Sayer, C. et al. The role of pond management for biodiversity conservation in an agricultural landscape. Aquat. Conserv. 22, 626-638 (2012).</td><td></td><td></td><td></td><td></td><td></td><td></td><td>√</td><td></td><td></td><td></td><td></td><td>√</td></tr>
<tr><td>9</td><td>Thornhill, I. A. et al. The functional response and resilience in small waterbodies along land-use and environmental gradients. Global. Change Biol. 24, 3079-3092 (2018).</td><td></td><td></td><td></td><td></td><td>√</td><td>√</td><td>√</td><td>√</td><td></td><td></td><td></td><td>√</td></tr>
<tr><td>10</td><td>Zhang, Y. et al. The potential benefits of on-farm mitigation scenarios for reducing multiple pollutant loadings in prioritized agri-environment areas across England. Environ. Sci. Policy 73, 100-114 (2017).</td><td></td><td></td><td></td><td></td><td>√</td><td>√</td><td></td><td>√</td><td></td><td>√</td><td></td><td></td></tr>
<tr><td rowspan="5">15</td><td>1</td><td>Apinda, L. E. A. et al. Value of artificial ponds for aquatic beetle and bug conservation in the Cape Floristic Region biodiversity hotspot. Aquat. Conserv. 24, 522-535 (2014).</td><td></td><td></td><td></td><td></td><td></td><td></td><td>√</td><td></td><td></td><td></td><td></td><td>√</td></tr>
<tr><td>2</td><td>Carta, A. Seed regeneration in Mediterranean temporary ponds: Germination ecophysiology and vegetation processes. Hydrobiologia, 782, 23-35 (2016).</td><td></td><td></td><td>√</td><td></td><td></td><td></td><td>√</td><td></td><td></td><td></td><td></td><td>√</td></tr>
<tr><td>3</td><td>De Necker, L. et al. Using stable δ13C and δ15N isotopes to assess food web structures in an African subtropical temporary pool. Afr. Zool. 55, 79-92 (2020).</td><td></td><td></td><td></td><td></td><td>√</td><td></td><td>√</td><td></td><td></td><td></td><td></td><td>√</td></tr>
<tr><td>4</td><td>Melly, B. L. et al. Mapping ephemeral wetlands: Manual digitization and logistic regression modelling in Nelson Mandela Bay Municipality, South Africa. Wetl. Ecol. Manag. 25, 313-330 (2017).</td><td>√</td><td></td><td></td><td></td><td></td><td></td><td>√</td><td></td><td></td><td>√</td><td></td><td>√</td></tr>
<tr><td>5</td><td>Namugize, J. N. et al. Effects of land use and land cover changes on water quality in the umngeni river catchment, South Africa. Phys. Chem. Earth. 105, 247-264 (2018).</td><td></td><td></td><td></td><td></td><td>√</td><td></td><td></td><td></td><td></td><td></td><td></td><td></td></tr>
</table>

Site ID	Id	Article	HM	WR	HP	BP	NR	GG	BD	ES	SA	IM	CC	WC
15	6	Rockstrom, J. Water resources management in smallholder farms in Eastern and Southern Africa: An overview. Phys. Chem. Earth. 25, 275-283 (2000).		√										√
	7	Samways, M. J. et al. Value of artificial ponds for aquatic insects in drought-prone southern Africa: A review. Biodivers. Conserv. 29, 3131-3150 (2020).		√					√				√	√
	8	Vimercati, G. et al. Integrating age structured and landscape resistance models to disentangle invasion dynamics of a pond-breeding anuran. Ecol. Model. 356, 104-116 (2017).						√	√					
16	1	Behera, M. D. et al. Wetland monitoring, serving as an index of land use change-A study in Samaspur wetlands, Uttar Pradesh, India. J. Indian. Soc. Remote. 40, 287-297 (2012).							√			√		√
	2	Rahman, M. M. et al. An enhanced swat wetland module to quantify hydraulic interactions between riparian depressional wetlands, rivers and aquifers. Environ. Modell. Softw. 84, 263-289 (2016).	√	√	√									√
	3	Singh, E. J. et al. Groundwater quality in Imphal West district, Manipur, India, with multivariate statistical analysis of data. Environ. Sci. Pollut. R. 20, 2421-2434 (2013).		√		√					√			
	4	Singh, M. and Sinha, R. Distribution, diversity, and geomorphic evolution of floodplain wetlands and wetland complexes in the Ganga plains of north Bihar, India. Geomorphology 351, 106960 (2020).	√		√							√		
	5	Sundar, K. G. and Kittur, S. Can wetlands maintained for human use also help conserve biodiversity? Landscape-scale patterns of bird use of wetlands in an agricultural landscape in north India. Biol. Conserv. 168, 49-56 (2013).		√					√		√	√		√
	6	Urfi, A. J. et al. Nesting ecology of the painted stork mycteria leucocephala at Sultanpur National Park, Haryana, India. Forktail 23, 150 (2007).							√					√
17	1	Fan, F. et al. Land use and land cover change in Guangzhou, China, from 1998 to 2003, based on Landsat TM/ETM+ imagery. Sensors 7, 1323-1342 (2007).										√		
	2	Hu, Y. et al. Environmental and human health challenges of industrial livestock and poultry farming in China and their mitigation. Environ. Int. 107, 111-130 (2017).		√		√	√	√			√		√	
	3	Jiang, J. et al. A modeling approach to evaluating the impacts of policy-induced land management practices on non-point source pollution: A case study of the Liuxi River watershed, China. Agr. Water Manage. 131, 1-16 (2014).			√		√				√			
	4	Li, F. et al. Analyzing trends of dike-ponds between 1978 and 2016 using multi-source remote sensing images in Shunde district of south China. Sustainability 10, 3504 (2018).		√							√	√		√
	5	Lu, J. and Li, X. Review of rice-fish-farming systems in China-One of the globally important ingenious agricultural heritage systems (GIAHS). Aquaculture 260, 106-113 (2006).					√	√	√		√		√	√
	6	Tang, Z. Rural revitalization and scientific management in the Pearl River Delta-Scientific decision based on scientific rationality and public understanding. Global Transitions 1, 241-250 (2019).									√			
	7	Weng, Q. A historical perspective of river basin management in the Pearl River Delta of China. J. Environ. Manage. 85, 1048-1062 (2007).	√	√							√			√

Continued

Site ID	Id	Article	HM	WR	HP	BP	NR	GG	BD	ES	SA	IM	CC	WC
17	8	Xiao, D. N. Landscape ecological construction in rural China: Theory and application. Chinese. Geogr. Sci. 11, 104-114 (2001).			√					√	√			√
18	1	Chen, H. Surface-flow constructed treatment wetlands for pollutant removal: Applications and perspectives. Wetlands 31, 805-814 (2011).	√	√	√	√			√	√				√
	2	Chen, W. et al. Exploring the multi-scale hydrologic regulation of multi-pond systems in a humid agricultural catchment. Water Res. 184, 115987 (2020).	√	√			√			√				√
	3	Fang, T. et al. Study on the application of integrated eco-engineering in purifying eutrophic river waters. Ecol. Eng. 94, 320-328 (2016).				√	√							
	4	Liu, Y. et al. Phosphorus sorption and sedimentation in a multi-pond system within a headstream agricultural watershed. Water Qual. Res. J. Can. 44, 243-252 (2009).				√	√							
	5	Tang, W. et al. Phosphorus buildup and release risk associated with agricultural intensification in the estuarine sediments of Chaohu Lake valley, eastern China. Clean-Soil Air Water 38, 336-343 (2010).				√	√				√			
	6	Verhoeven, J. T. et al. Regional and global concerns over wetlands and water quality. Trends Ecol. Evol. 21, 96-103 (2006).	√	√	√	√	√	√	√	√				√
	7	Wu, D. et al. Improvement and testing of SWAT for multi-source irrigation systems with paddy rice. J. Hydrol. 568, 1031-1041 (2019).	√	√	√									
	8	Xia, Y. et al. Is indirect N2O emission a significant contributor to the agricultural greenhouse gas budget? A case study of a rice paddy-dominated agricultural watershed in eastern China. Atmos. Environ. 77, 943-950 (2013).				√	√	√			√		√	
	9	Yin, C. and Shan, B. Multi-pond systems: A sustainable way to control diffuse phosphorus pollution. Ambio 30, 369-375 (2001).	√	√	√	√			√					
	10	Zhang, H. and Shan, B. Historical distribution and partitioning of phosphorus in sediments in an agricultural watershed in the Yangtze-Huaihe region, China. Environ. Sci. Technol. 42, 2328-2333 (2008).				√	√							
	11	Zhang, H. and Shan, B. Historical records of heavy metal accumulation in sediments and the relationship with agricultural intensification in the Yangtze-Huaihe region, China. Sci. Total. Environ. 399, 113-120 (2008).				√	√							
19	1	Akasaka, M. et al. Effects of land use on aquatic macrophyte diversity and water quality of ponds. Freshwater Biol. 55, 909-922 (2010).					√		√					√
	2	Deguchi, S. et al. Ponds support higher bird diversity than rice paddies in a hilly agricultural area in Japan. Biodivers. Conserv. 29, 3265-3285 (2020).							√					√
	3	Hamasaki, K. et al. Relative importance of within-habitat environment, land use and spatial autocorrelations for determining odonate assemblages in rural reservoir ponds in Japan. Ecol. Res. 24, 597-605 (2009).							√					√
	4	Kasahara, S. et al. Rice fields along the East Asian-Australasian flyway are important habitats for an inland wader's migration. Sci. Rep-UK 10, 1-9 (2020).							√	√	√		√	√
	5	Maezono, Y. and Miyashita, T. Community-level impacts induced by introduced largemouth bass and bluegill in farm ponds in Japan. Biol. Conserv. 109, 111-121 (2003).							√					

Site ID	Id	Article	HM	WR	HP	BP	NR	GG	BD	ES	SA	IM	CC	WC
19	6	Natuhara, Y. Ecosystem services by paddy fields as substitutes of natural wetlands in Japan. Ecol. Eng. 56, 97-106 (2013).			√	√		√	√	√				√
	7	Usio, N. et al. Effects of land use on trophic states and multi-taxonomic diversity in Japanese farm ponds. Agr. Ecosyst. Environ. 247, 205-215 (2017).							√	√				√
	8	Usio, N. et al. Effects of pond draining on biodiversity and water quality of farm ponds. Conserv. Biol. 27, 1429-1438 (2013).				√			√	√				√
	9	Wang, Z. et al. Retrieval of chlorophyll-a and total suspended solids using iterative stepwise elimination partial least squares (ISE-PLS) regression based on field hyperspectral measurements in irrigation ponds in Higashihiroshima, Japan. Remote. Sens. 9, 264 (2017).				√								
20	1	Cartwright, I. and Morgenstern, U. Using tritium to document the mean transit time and sources of water contributing to a chain-of-ponds river system: implications for resource protection. Appli. Geochem. 75, 9-19 (2016).		√	√									√
	2	Hazell, D. et al. A comparison of constructed and natural habitat for frog conservation in an Australian agricultural landscape. Biol. Conserv. 119, 61-71 (2004).							√					√
	3	Mactaggart, B. et al. When History May Lead us Astray: using historical documents to reconstruct swampy meadows/chains of ponds in the New South Wales Central Tablelands, Australia. Aust. Geogr. 38, 233-252 (2007).	√									√		√
	4	Mould, S. and Fryirs, K. The Holocene evolution and geomorphology of a chain of ponds, southeast Australia: Establishing a physical template for river management. Catena 149, 349-362 (2017).	√											√
	5	Williams, R. T. and Fryirs, K. A. The morphology and geomorphic evolution of a large chain-of-ponds river system. Earth Surf. Proc. Land. 45, 1732-1748 (2020).	√											√
	6	Williams, R. T. et al. The hydrological function of a large chain-of-ponds: A wetland system with intermittent surface flows. Aquat. Sci. 82, 1-18 (2020).	√		√									
21	1	Bennett, J. R. et al. Polar lessons learned: Long-term management based on shared threats in Arctic and Antarctic environments. Front. Ecol. Environ. 13, 316-324 (2015).				√		√	√				√	√
	2	Bowling, L. C. and Lettenmaier, D. P. Modeling the effects of lakes and wetlands on the water balance of Arctic environments. J. Hydrometeorol. 11, 276-295 (2010).			√									
	3	Clewley, D. et al. Evaluation of ALOS PALSAR data for high-resolution mapping of vegetated wetlands in Alaska. Remote. Sens. 7, 7272-7297 (2015).						√				√	√	
	4	Hinzman, L. D. et al. Evidence and implications of recent climate change in northern Alaska and other arctic regions. Climatic Change 72, 251-298 (2005).	√		√	√		√					√	
	5	In't Zandt, M. H., Liebner, S., and Welte, C. U. Roles of thermokarst lakes in a warming world. Trends. Microbiol. 28, 769-779 (2020).				√		√					√	
	6	Kallistova, A. Y. et al. Thermokarst lakes, ecosystems with intense microbial processes of the methane cycle. Microbiology+ 88, 649-661 (2019).			√	√		√	√					

Continued

Site ID	Id	Article	HM	WR	HP	BP	NR	GG	BD	ES	SA	IM	CC	WC
21	7	Klein, E. et al. Wetland drying and succession across the Kenai Peninsula Lowlands, south-central Alaska. Can. J. Forest. Res. 35, 1931-1941 (2005).		√									√	
	8	Necsoiu, M. et al. Multi-temporal image analysis of historical aerial photographs and recent satellite imagery reveals evolution of water body surface area and polygonal terrain morphology in Kobuk Valley National Park, Alaska. Environ. Res. Lett. 8, p.025007 (2013).	√		√					√		√	√	
	9	Rains, M. C. Water sources and hydrodynamics of closed-basin depressions, cook inlet region, Alaska. Wetlands 31, 377-387 (2011).			√	√							√	
	10	Rautio, M. et al. Shallow freshwater ecosystems of the circumpolar Arctic. Ecoscience 18, 204-222 (2011).	√		√			√	√				√	
	11	Riordan, B. et al. Shrinking ponds in subarctic Alaska based on 1950-2002 remotely sensed images. J. Geophys. Res-Biogeo. 111, (2006).			√					√				
	12	Schmidt, J. H. et al. Environmental and human influences on trumpeter swan habitat occupancy in Alaska. Condor. 111, 266-275 (2009).							√				√	
	13	Tiegs, S. D. et al. Litter decomposition, and associated invertebrate communities, in wetland ponds of the Copper River Delta, Alaska (USA). Wetlands 33, 1151-1163 (2013).					√	√	√	√				√
	14	Wik, M. et al. Climate-sensitive northern lakes and ponds are critical components of methane release. Nat. Geosci. 9, 99-105 (2016).							√					
	15	Van Huissteden, J., and Dolman, A. J. Soil carbon in the Arctic and the permafrost carbon feedback. Curr. Opin. Env. Sust. 4, 545-551 (2012).	√		√	√		√					√	
	16	Yoshikawa, K. and Hinzman, L. D. Shrinking thermokarst ponds and groundwater dynamics in discontinuous permafrost near Council, Alaska. Permafrost. Periglac. 14, 151-160 (2003).	√		√									
22	1	Ameli, A. A. and Creed, I. F. Does wetland location matter when managing wetlands for watershed-scale flood and drought resilience? J. Am. Water Resour. As. 55, 529-542 (2019).	√		√									√
	2	Bartzen, B. A. et al. Trends in agricultural impact and recovery of wetlands in prairie Canada. Ecol. Appl. 20, 525-538 (2010).							√	√				√
	3	Benoy, G. A. Tiger salamanders in prairie potholes: A "fish in amphibian's garments?". Wetlands 28, 464-472 (2008).							√					
	4	Brooks, J. R. et al. Estimating wetland connectivity to streams in the prairie pothole region: An isotopic and remote sensing approach. Water Resour. Res. 54, 955-977 (2018).		√	√	√						√		
	5	Evenson, G. R. et al. Depressional wetlands affect watershed hydrological, biogeochemical, and ecological functions. Ecol. Appl. 28, 953-966 (2018).	√	√	√	√	√		√	√				
	6	Fedy, B. C. et al. Distribution of priority grassland bird habitats in the Prairie Pothole Region of Canada. Avian. Conserv. Ecol. 13, 4 (2018).							√					√
	7	Forcey, G. M. et al. Influence of land use and climate on wetland breeding birds in the Prairie Pothole region of Canada. Can. J. Zool. 85, 421-436 (2007).							√					
	8	Muhammad, A. et al. Assessing the importance of potholes in the Canadian prairie region under future climate change scenarios. Water 10, 1657 (2018).		√	√			√		√			√	

Continued

Site ID	Id	Article	HM	WR	HP	BP	NR	GG	BD	ES	SA	IM	CC	WC
22	9	Neff, B. P. and Rosenberry, D. O. Groundwater connectivity of upland-embedded wetlands in the prairie pothole region. Wetlands 38, 51-63 (2018).			√									
	10	Pattison, J. K. et al. Wetlands, flood control and ecosystem services in the Smith Creek Drainage Basin: A case study in Saskatchewan, Canada. Ecol. Econ. 147, 36-47 (2018).			√				√	√			√	√
	11	Shaw, D. A. et al. Topographic analysis for the prairie pothole region of western Canada. Hydrol. Process. 27, 3105-3114 (2013).	√	√	√									
	12	van der Valk, A. G. and Pederson, R. L. The SWANCC decision and its implications for prairie potholes. Wetlands 23, 590-596 (2003).				√		√	√	√				√
	13	Wu, Q. et al. Integrating LiDAR data and multi-temporal aerial imagery to map wetland inundation dynamics using Google earth engine. Remote. Sens. Environ. 228, 1-13 (2019).	√	√	√							√		
	14	Zeng, T. and Arnold, W. A. Pesticide photolysis in prairie potholes: probing photosensitized processes. Environ. Sci. Technol. 47, 6735-6745 (2013).				√	√							
23	1	Delidjakova, K. K. et al. Influence of Hudson Bay on the carbon dynamics of a Hudson bay lowlands coastal site. Arct. Sci. 2, 142-163 (2016).				√		√						
	2	Glaser, P. H. et al. Rates, pathways and drivers for peatland development in the Hudson bay lowlands, northern Ontario, Canada. J. Ecol. 92, 1036-1053 (2004).	√		√									
	3	Harris, L. I. et al. Mechanisms for the development of microform patterns in peatlands of the Hudson bay lowland. Ecosystems 23, 741-767 (2020).	√		√	√	√		√					
	4	Helbig, M. et al. Contrasting temperature sensitivity of CO2 exchange in peatlands of the Hudson bay lowlands, Canada. J. Geophys. Res-Biogeo. 124, 2126-2143 (2019).				√	√	√					√	
	5	Holmquist, J. R. et al. Peatland initiation, carbon accumulation, and 2 ka depth in the James bay lowland and adjacent regions. Arct. Antarct. Alp. Res. 46, 19-39 (2014).				√		√					√	
	6	Humphreys, E. R. et al. Two bogs in the Canadian Hudson bay lowlands and a temperate bog reveal similar annual net ecosystem exchange of CO2. Arct. Antarct. Alp. Res. 46, 103-113 (2014).				√							√	
	7	Laamrani, A. et al. Analysis of the effect of climate warming on paludification processes: Will soil conditions limit the adaptation of Northern Boreal Forests to climate change? A Synthesis. Forests 11, 1176 (2020).				√		√	√				√	
	8	Martini, I. P. The cold-climate peatlands of the Hudson Bay Lowland, Canada: Brief overview of recent work. Developments in Earth Surface Processes 9, 53-84 (2006).	√			√		√	√					
	9	Richardson, M. et al. The influences of catchment geomorphology and scale on runoff generation in a northern peatland complex. Hydrol. Process. 26, 1805-1817 (2012).	√		√								√	
	10	Ulanowski, T. A. and Branfireun, B. A. Small-scale variability in peatland pore-water biogeochemistry, Hudson Bay Lowland, Canada. Sci. Total Environ. 454, 211-218 (2013).				√							√	

Continued

Site ID	Id	Article	HM	WR	HP	BP	NR	GG	BD	ES	SA	IM	CC	WC
24	1	Åhlén, I. et al. Wetlandscape size thresholds for ecosystem service delivery: Evidence from the Norrström drainage basin, Sweden. Sci. Total. Environ. 704, 135452 (2020).			√		√		√	√				√
	2	Arheimer, B. and Pers, B. C. Lessons learned? Effects of nutrient reductions from constructing wetlands in 1996-2006 across Sweden. Ecol. Eng. 103, 404-414 (2017).					√			√	√			
	3	Arheimer, B. and Wittgren, H. Modelling nitrogen removal in potential wetlands at the catchment scale. Ecol. Eng. 19, 63-80 (2002).			√	√	√							
	4	Bring, A. et al. Groundwater storage effects from restoring, constructing or draining wetlands in temperate and boreal climates: A systematic review protocol. Environ. Evid. 9, 1-11 (2020).		√	√		√	√						
	5	Darracq, A. et al. Quantification of advective solute travel times and mass transport through hydrological catchments. Environ. Fluid Mech. 10, 103-120 (2010).			√	√	√							
	6	Darracq, A. et al. Scale and model resolution effects on the distributions of advective solute travel times in catchments. Hydrol. Process. 24, 1697-1710 (2010).			√	√	√							
	7	Destouni, G. et al. General quantification of catchment-scale nutrient and pollutant transport through the subsurface to surface and coastal waters. Environ. Sci. Technol. 44, 2048-2055 (2010).					√	√						
	8	Elmhagen, B. et al. Interacting effects of change in climate, human population, land use, and water use on biodiversity and ecosystem services. Ecol. Soc. 20, 23 (2015).						√	√	√			√	
	9	Moor, H. et al. Predicting climate change effects on wetland ecosystem services using species distribution modeling and plant functional traits. Ambio 44, 113-126 (2015).			√	√	√	√	√	√			√	√
	10	Persson, K. et al. Diffuse hydrological mass transport through catchments: scenario analysis of coupled physical and biogeochemical uncertainty effects. Hydrol. Earth Syst. Sc. 15, 3195-3206 (2011).			√	√	√							
	11	Quin, A. et al. Dissecting the ecosystem service of large-scale pollutant retention: The role of wetlands and other landscape features. Ambio 44, 127-137 (2015).			√	√	√			√				√
25	1	Bukvareva, E. N. et al. The current state of knowledge of ecosystems and ecosystem services in Russia: A status report. Ambio 44, 491-507 (2015).			√			√	√	√			√	√
	2	Kasimov, N. S. et al. Modern geochemical evolution of lagoon-marshy landscapes in the western Caspian Sea region. Eurasian Soil. Sci. 45, 1-11 (2012).	√		√								√	
	3	Kholodov, V. N. et al. Facies types of sedimentary iron ore deposits and their geochemical features: Communication 1. Facies groups of sedimentary ores, their lithology, and genesis. Lithol. Miner. Resour. 47, 447-472 (2012).			√									
	4	Lychagin, M. Y. et al. Heavy metals in the water, plants, and bottom sediments of the Volga River mouth area. J. Coastal. Res. 31, 859-868 (2015).	√		√									
	5	Robarts, R. D. et al. The state of knowledge about wetlands and their future under aspects of global climate change: The situation in Russia. Aquat. Sci. 75, 27-38 (2013).		√	√	√				√			√	√
	6	Tanneberger, F. et al. The peatland map of Europe. Mires. Peat. 19, 1-17 (2017).							√	√		√	√	

Site ID	Id	Article	HM	WR	HP	BP	NR	GG	BD	ES	SA	IM	CC	WC
26	1	Fonseca, A. et al. Integrated hydrological and water quality model for river management: A case study on Lena River. Sci. Total. Environ. 485, 474-489 (2014).			√									
	2	Helbig, M. et al. Spatial and seasonal variability of polygonal tundra water balance: Lena River Delta, northern Siberia (Russia). Hydrogeol. J. 21, 133-147 (2013).	√	√	√									
	3	Razjigaeva, N. G. et al. Landscape response to the medieval warm period in the south Russian far east. Quatern. Int. 519, 215-231 (2019).							√					
	4	Suzuki, K. et al. Satellite gravimetry-based analysis of terrestrial water storage and its relationship with run-off from the Lena River in eastern Siberia. Int. J. of Remote. Sens. 37, 2198-2210 (2016).		√										
	5	Vompersky, S. E. et al. Estimation of forest cover extent over peatlands and paludified shallow-peat lands in Russia. Contemp. Probl. Ecol. 4, 734-741 (2011).	√					√	√			√		
27	1	Krogulec, E. et al. Hydrogeological characteristics of aquifer near Arctowski Polish Antarctic station on King George Island (South Shetland Islands), Antarctica. Polar. Sci. 16, 68-77 (2018).												√
	2	Kvíderová, J. and Elster, J. Standardized algal growth potential and/or algal primary production rates of maritime Antarctic stream waters (King George Island, South Shetlands). Polar. Res. 32, 11191 (2013).							√					
	3	Šabacká, M. and Elster, J. Response of cyanobacteria and algae from Antarctic wetland habitats to freezing and desiccation stress. Polar Biol. 30, 31-37 (2006).							√					
28	1	Christiansen, J. R. et al. Methane fluxes and the functional groups of methanotrophs and methanogens in a young Arctic landscape on Disko Island, West Greenland. Biogeochemistry 122, 15-33 (2015).				√								
	2	Joabsson, A. and Christensen, T. R. Methane emissions from wetlands and their relationship with vascular plants: an Arctic example. Global. Change. Biol. 7, 919-932 (2001).				√		√	√					
	3	Pirk, N. et al. Toward a statistical description of methane emissions from arctic wetlands. Ambio 46, 70-80 (2017).				√		√				√		
	4	Schuldt, R. J. et al. Modelling Holocene carbon accumulation and methane emissions of boreal wetlands—an Earth system model approach. Biogeosciences 10, 1659-1674 (2013).				√		√				√		
	5	Ström, L. et al. Presence of Eriophorum scheuchzeri enhances substrate availability and methane emission in an Arctic wetland. Soil. Biol. Biochem. 45, 61-70 (2012).				√		√	√					
	6	Westergaard, A. et al. Camera derived vegetation greenness index as proxy for gross primary production in a low Arctic wetland area. Isprs. J. Photogramm. 86, 89-99 (2013).							√	√		√		
	7	Woo, M. K. and Young, K. L. High Arctic wetlands: their occurrence, hydrological characteristics and sustainability. J. Hydrol. 320, 432-450 (2006).	√		√								√	
29	1	Gao, J. et al. Topographic influence on wetland distribution and change in Maduo county, Qinghai-Tibet Plateau, China. J. Mt. Sci-Engl. 9, 362-371 (2012).	√	√								√		

Continued

Site ID	Id	Article	HM	WR	HP	BP	NR	GG	BD	ES	SA	IM	CC	WC
29	2	Gao, J. et al. Degradation of wetlands on the Qinghai-Tibet Plateau: A comparison of the effectiveness of three indicators. J. Mt. Sci-Engl. 10, 658-667 (2013).			√			√		√				
	3	Gao, J. et al. Geomorphic-centered classification of wetlands on the Qinghai-Tibet Plateau, Western China. J. Mt. Sci-Engl. 10, 632-642 (2013).	√	√										√
	4	Jin, H. J. et al. Changes in permafrost environments along the Qinghai–Tibet engineering corridor induced by anthropogenic activities and climate warming. Cold. Reg. Sci. Technol. 53, 317-333 (2008).			√								√	
	5	Şerban, R. D. et al. Mapping thermokarst lakes and ponds across permafrost landscapes in the headwater area of yellow river on northeastern Qinghai-Tibet plateau. Int. J. Remote. Sens. 41, 7042-7067 (2020).	√	√	√							√	√	
	6	Wang, J. et al. High uncertainties detected in the wetlands distribution of the Qinghai–Tibet Plateau based on multisource data. Landsc. Ecol. Eng. 16, 47-61 (2020).						√	√	√		√	√	√
	7	Zhang, L. et al. Significant methane ebullition from alpine permafrost rivers on the east Qinghai-Tibet plateau. Nat. Geosci. 13, 349-354 (2020).				√		√					√	
30	1	Head, J. W. and Marchant, D. R. The climate history of early Mars: Insights from the Antarctic Mcmurdo dry valleys hydrologic system. Antarct. Sci. 26, 774-800 (2014).	√		√	√		√					√	
	2	Power, S. N. et al. Estimating microbial mat biomass in the Mcmurdo dry valleys, Antarctica using satellite imagery and ground surveys. Polar. Biol. 43, 1753-1767 (2020).	√			√				√				√
	3	Jungblut, A. D. et al. The Pyramid Trough Wetland: environmental and biological diversity in a newly created Antarctic protected area. Fems. Microbiol. Ecol. 82, 356-366 (2012).				√				√				√
	4	Sutherland, D. L. et al. Environmental drivers that influence microalgal species in meltwater pools on the Mcmurdo ice shelf, Antarctica. Polar. Biol. 43, 467-482 (2020).				√				√			√	
	5	Wlostowski, A. N. et al. The hydroecology of an ephemeral wetland in the Mcmurdo dry valleys, Antarctica. J. Geophys. Res. 124, 3814-3830 (2019).			√	√				√				

[a] A tick mark indicates this research focus is included in the literature.

Table A9.7 Aggregated proportions (%) of research focus class for the 30 representative NFW regions, categorized according to the five climate zones used.

Climate zone	HM	WR	HP	BP	NR	GG	BD	ES	SA	IM	CC	WC
Tropical	10.26-[a]	17.95-	38.46 +	30.77	7.69-	51.28 +	46.15 +	38.46 +	12.82-	20.51	51.28 +	48.72 +
Arid	24.32	21.62	56.76 +	43.24 +	5.41-	2.70-	48.65 +	27.03	18.92-	8.11-	18.92-	48.65 +
Temperate	15.94-	26.09	26.09	31.88	30.43	10.14-	44.93 +	23.19	26.09	13.04-	14.49-	57.97 +
Cold	27.42	17.74-	53.23 +	45.16 +	22.58	32.26	35.48 +	24.19	4.84-	9.68-	38.71 +	19.35
Polar	27.27	13.64-	27.27	50.00 +	0.00-	36.36 +	45.45 +	13.64-	0.00-	18.18-	40.91 +	22.73

[a] The plus and minus marks indicate the three groups of all proportional data classified by the Jenks classification method implemented in ArcGIS.

References

Åhlén, I., Vigouroux, G., Destouni, G., et al., 2021. "Hydro-climatic changes of wetlandscapes across the world." *Sci. Rep.* 11: 1-11.

Bertassello, L. E., Jawitz, J. W., Aubeneau, A. F., et al., 2019. "Stochastic dynamics of wetlandscapes: Ecohydrological implications of shifts in hydro-climatic forcing and landscape configuration." *Sci. Total Env.* 694: 133765.

Jaramillo, F., Licero, L., Åhlen, I., et al., 2018. "Effects of hydroclimatic change and rehabilitation activities on salinity and mangroves in the Ciénaga Grande de Santa Marta, Colombia." *Wetlands.* 38(4): 755-767.

Jungblut, A. D., Wood, S. A., Hawes, I., et al., 2012. "The Pyramid Trough Wetland: Environmental and biological diversity in a newly created Antarctic protected area." *Fems. Microbiol. Ecol.* 82(2): 356-366.

Lane, C. R., and Amico, D. E., 2016. "Identification of putative geographically isolated wetlands of the conterminous United States." *J. Am. Water Resour. As.* 52(3): 705-722.

WHO, 2006. Ecosystems and human well-being: Health synthesis: A report of the Millennium Ecosystem Assessment. Geneva: WHO Retrieved from.

Wlostowski, A. N., Gooseff, M. N., McKnight, D. M., et al., 2016. "Patterns of hydrologic connectivity in the McMurdo Dry Valleys, Antarctica: A synthesis of 20 years of hydrologic data." *Hydrol. Process.* 30(17): 2958-2975.